Classical Competing RISKS

Classical Competing RISKS

Martin J. Crowder

CHAPMAN & HALL/CRC

Boca Raton London New York Washington, D.C.

Library of Congress Cataloging-in-Publication Data

Crowder, M. J. (Martin J.), 1943-
Classical competing risks / Martin John Crowder.
p. cm.
Includes bibliographical references and index.
ISBN 1-58488-175-5 (alk. paper)
1. Competing risks. 2. Failure time data analysis. I. Title.

QA273 .C855 2001
519.2--dc21 2001025012
CIP

Direct all inquiries to CRC Press LLC, 2000 N.W. Corporate Blvd., Boca Raton, Florida 33431.

International Standard Book Number 1-58488-175-5
Library of Congress Card Number 2001025012
Printed in the United States of America 1 2 3 4 5 6 7 8 9 0
Printed on acid-free paper

For

Charles Stanley Crowder

Contents

Chapter 1 Continuous failure times and their causes ... 1
1.1 Basic probability functions ... 1
1.1.1 Univariate survival distributions ... 1
1.1.2 A time and a cause: Competing Risks ... 4
1.2 Some small data sets ... 6
1.3 Hazard functions ... 10
1.3.1 Sub-hazards and overall hazard ... 10
1.3.2 Proportional hazards ... 12
1.4 Regression models ... 13
1.4.1 Proportional hazards ... 15
1.4.2 Accelerated life ... 16
1.4.3 Proportional odds ... 16
1.4.4 Mean residual life ... 16

Chapter 2 Parametric likelihood inference ... 19
2.1 The likelihood for Competing Risks ... 19
2.1.1 Forms of the likelihood function ... 19
2.1.2 Incomplete observation of C or T ... 20
2.1.3 Maximum likelihood estimates ... 21
2.2 Model checking ... 22
2.2.1 Goodness of fit ... 22
2.2.2 Residuals ... 23
2.3 Inference ... 24
2.3.1 Hypothesis tests and confidence intervals ... 24
2.3.2 Bayesian approach ... 24
2.3.3 Bayesian computation ... 26
2.4 Some examples ... 27
2.5 Masked systems ... 34

Chapter 3 Latent failure times: probability distributions ... 37
3.1 Basic probability functions ... 37
3.1.1 Multivariate survival distributions ... 37
3.1.2 Latent lifetimes ... 38
3.2 Some examples ... 40
3.3 Marginal vs. sub-distributions ... 46

3.4 Independent risks ... 48
3.4.1 The Makeham assumption ... 50
3.4.2 Proportional hazards ... 51
3.4.3 Some examples ... 52
3.5 A risk-removal model ... 53

Chapter 4 Likelihood functions for univariate survival data ... 57
4.1 Discrete and continuous failure times ... 57
4.1.1 Discrete survival distributions ... 58
4.1.2 Mixed survival distributions ... 59
4.2 Discrete failure times: estimation ... 62
4.2.1 Random samples: parametric estimation ... 62
4.2.2 Random samples: non-parametric estimation ... 63
4.2.3 Explanatory variables ... 65
4.2.4 Interval-censored data ... 66
4.3 Continuous failure times: random samples ... 67
4.3.1 The Kaplan-Meier estimator ... 67
4.3.2 The integrated and cumulative hazard functions ... 68
4.4 Continuous failure times: explanatory variables ... 69
4.4.1 Cox's proportional hazards model ... 69
4.4.2 Cox's partial likelihood ... 70
4.4.3 The baseline survivor function ... 72
4.4.4 Residuals ... 73
4.4.5 Some theory for partial likelihood ... 74
4.5 Discrete failure times again ... 76
4.5.1 Proportional hazards ... 76
4.5.2 Proportional odds ... 77
4.6 Time-dependent covariates ... 78

Chapter 5 Discrete failure times in Competing Risks ... 83
5.1 Basic probability functions ... 83
5.2 Latent failure times ... 85
5.3 Some examples based on Bernoulli trials ... 89
5.4 Likelihood functions ... 92
5.4.1 Parametric likelihood ... 93
5.4.2 Non-parametric estimation from random samples ... 94
5.4.3 Asymptotic distribution of non-parametric estimators ... 97

Chapter 6 Hazard-based methods for continuous failure times ... 101
6.1 Latent failure times vs. hazard modelling ... 101
6.2 Some examples of hazard modelling ... 102
6.3 Non-parametric methods for random samples ... 106
6.3.1 The Kaplan-Meier estimator ... 106
6.3.2 Interval-censored data ... 109
6.3.3 Actuarial approach ... 111

6.4 Proportional hazards and partial likelihood ... 113
6.4.1 The proportional hazards model ... 113
6.4.2 The partial likelihood ... 114
6.4.3 The baseline survivor functions ... 116
6.4.4 Discrete failure times ... 117

Chapter 7 Latent failure times: identifiability crises ... 119
7.1 The Cox-Tsiatis impasse ... 119
7.2 More general identifiablility results ... 122
7.3 Specified marginals ... 130
7.4 Discrete failure times ... 134
7.5 Regression case ... 137
7.6 Censoring of survival data ... 139
7.7 Parametric identifiability ... 141

Chapter 8 Martingale counting processes in survival data ... 143
8.1 Introduction ... 143
8.2 Back to basics: probability spaces and conditional expectation ... 144
8.3 Filtrations ... 147
8.4 Martingales ... 148
8.4.1 Discrete time ... 148
8.4.2 Continuous time ... 150
8.5 Counting processes ... 151
8.6 Product integrals ... 153
8.7 Survival data ... 154
8.7.1 A single lifetime ... 154
8.7.2 Independent lifetimes ... 155
8.7.3 Competing risks ... 157
8.7.4 Right-censoring ... 158
8.8 Non-parametric estimation ... 159
8.8.1 Survival times ... 159
8.8.2 Competing risks ... 160
8.9 Non-parametric testing ... 161
8.10 Regression models ... 163
8.10.1 Intensity models and time-dependent covariates ... 163
8.10.2 Proportional hazards model ... 164
8.10.3 Martingale residuals ... 164
8.11 Epilogue ... 165

Appendix 1 Numerical maximisation of likelihood functions ... 167

Appendix 2 Bayesian computation ... 171

Bibliography ... 173

Index ... 183

Preface

If something can fail, it can often fail in one of several ways and sometimes in more than one way at a time. This phenomenon is amply illustrated by the MOT (Ministry of Transport) testing records of my cars and bikes over the years. One thing that has long puzzled me is just how much work in Survival Analysis has been accomplished without mentioning Competing Risks. There is always some cause of failure and almost always more than one possible cause. In one sense, then, Survival Analysis is a lost cause.

The origins of Competing Risks can be traced back to Daniel Bernoulli's attempt in 1760 to disentangle the risk of dying from smallpox from other risks. Seal (1977) and David and Moeschberger (1977, Appendix A) provide historical accounts. Much of the work in the subject since that time has been demographic and actuarial in nature. However, the results of major importance for statistical inference, and applications of the theory in Reliability, are quite recent, the two fields themselves being relatively recent.

One problem with writing a book on a specialized topic is that one has to start somewhere. Different readers will have differing levels of prior knowledge. Too much or too little assumed can be disastrous, and only "just about right" will give the book a fighting chance of serious consideration. What all this is sidling up to is the decision to put certain foundational material into preliminary subsections in some chapters rather than to exclude it altogether or bury it in appendices. This decision was taken boldly and decisively, after many months of agonized dithering, shunting the stuff to and fro. Its final resting place will no doubt offend some, known as reviewers, but will be seen as a sensible choice by others, I hope. A related choice is to consider throughout only well-defined risks of real, objectively verifiable events: we leave it to the philosophers and psychologists to debate how many perceived risks can be balanced on a pinhead. Another problem with writing a book like this is that one has to stop somewhere. My stopping rule was, appropriately, at martingale methods used for the counting-process approach to Survival Analysis: this requires a discrete jump in the level of mathematical technicality. The last chapter gives a brief introduction, but then the reader who is inspired, and has not expired, is referred to weightier tomes to boldly go where this book has not gone before — hence the "Classical" in the title of this book. In short, the target level here is for readers with formal statistical training, experience in applied statistics, and familiarity with probability distributions, hazard functions, and Survival Analysis. Competing Risks is a specialized, but widely applicable, topic for the statistician. The book is probably not very suitable for scientists, engineers, biologists, etc. to dip into for a quick fix, nor for pure mathematicians interested only in technical theory.

The plan of the book is as follows. In Chapter 1 the bare bones of the Competing Risks set-up are laid out, so to speak, and the associated likelihood functions are presented in Chapter 2. Chapter 3 describes the approach to Competing Risks via a joint distribution of latent failure times. Chapter 4 covers more modern approaches to Survival Analysis via hazard functions as a preparation for the extension of these methods to Competing Risks. Chapter 5 carries this work on to discrete failure times in Competing Risks and Chapter 6 deals likewise with the continuous case. Chapter 7 reports some blockages along the traditional modelling route via latent failure times, in particular, difficulties of identifiability. Chapter 8 gives a brief introduction to the counting-process approach and the associated martingale theory.

Throughout, an attempt has been made to give credit where due for all results, examples, applications, etc. where known. I apologize if I have missed any innovators in this. The results are often not stated or proved in exactly the original way. This is not claimed to be an "improvement" on the originals — it just arises through trying to give a coherent, self-contained, sequential account of the subject. Nevertheless, there is the odd refinement: one can often strike more effectively after others have lit up the target. I have been influenced by many authors, though I hope that there is not too much brazen plagiarization here, at least not of the sort that would stand up in court. Standard works on which I have drawn include, in chronological order, the books by David and Moeschberger (1978), Kalbfleisch and Prentice (1980), Lawless (1982), Nelson (1982), and Cox and Oakes (1984) and the papers by Gail (1975) and Elandt-Johnson (1976). Equally excellent, more recent books which should be mentioned are those by Fleming and Harrington (1991), Andersen et al. (1993), Collett (1994), and Bedford and Cooke (2001).

My own involvement in the subject arose from being presented with engineering data and not liking to admit in public that I didn't have a clue what to do with it. Eventually, I found out and managed to use the methods in several diverse applications, in both engineering and medical contexts. However, real data always seem to have a sting in the tail in the sense that they are never quite straightforward enough to provide an effective illustration in a textbook. Thus, the experienced statistician will find that the examples here are presented rather cursorily. However, the experienced statistician will also understand that in this way the core subject can be focused upon without being drawn into side issues, important as they might be in the application context. What this means is that only the central theory and methodology is covered, leaving the practitioner to extend the treatment as necessary for his particular problems. But then, that's what makes the job of statisticians interesting and why we should resist our replacement by computer packages!

Martin John Crowder
Imperial College, London

chapter 1

Continuous failure times and their causes

The bare bones of the Competing Risks set-up are laid out in this chapter. All that is considered is a time and a cause, without the traditional, additional structure of latent failure times; the latter will be addressed in Chapter 3.

1.1 Basic probability functions

Before getting down to the serious business of Competing Risks in Section 1.1.2, we cover some preliminaries.

1.1.1 Univariate survival distributions

To the mighty probabilist, failure times are just positive, or non-negative, random variables, or measurable functions. To the rest of us, humble statisticians, they are the kind of data for which there has been a daunting development in methodology over the past 20 years, the result being a bewildering variety of techniques many of which are pretty hard to get to grips with at the technical level. This subsection is the first example of an attempt throughout this book to introduce things from the bottom up, i.e., to cover fairly basic terms and definitions, and give the odd simple example, before going on to apply them in the main context.

The **distribution function** F and the **survivor function** $\bar{F}$ of a random variable T are defined by

$$F(t) = \text{pr}(T \leq t), \; \bar{F}(t) = \text{pr}(T > t),$$

so $F(t) + \bar{F}(t) = 1$ for all t. Note that $F(t)$ is an increasing, and $\bar{F}(t)$ is a decreasing, function of t. For the time being, we will consider only continuous failure time variables. Then $F(0) = 0$ and $\bar{F}(0) = 1$, since T is naturally non-negative. The **density function** is defined as

$$f(t) = dF(t)/dt = -d\bar{F}(t)/dt;$$

correspondingly,

$$F(t) = \int_0^t f(s)ds, \quad \bar{F}(t) = \int_t^\infty f(s)ds.$$

Modern survival analysis is mostly based around hazard functions. These are concerned with the probability of imminent failure, i.e., that, having got this far, you will get no further. The formal definition of the **hazard function** h of T is

$$h(t) = \lim_{\delta \to 0} \delta^{-1} \mathrm{pr}(T \le t + \delta \mid T > t).$$

The right hand side is equal to

$$\lim_{\delta \to 0} \delta^{-1} \mathrm{pr}(t < T \le t + \delta)/\mathrm{pr}(T > t) = \lim_{\delta \to 0} \delta^{-1}\{\bar{F}(t) - \bar{F}(t + \delta)\}/\bar{F}(t)$$

$$= -\{d\bar{F}(t)/dt\}/\bar{F}(t) = -d\log\bar{F}(t)/dt.$$

In different contexts, $h(t)$ is variously known as the instantaneous failure rate, age-specific failure rate, age-specific death rate, intensity function, and force of mortality or decrement. Integration yields the inverse relationship

$$\bar{F}(t) = \exp\left\{-\int_0^t h(s)ds\right\} = \exp\{-H(t)\}, \tag{1.1.1}$$

where $H(t)$ is the **integrated hazard function** and the lower limit 0 of the integral is consistent with $\bar{F}(0) = 1$. For a proper lifetime distribution, i.e., one for which $\bar{F}(\infty) = 0$, $H(t)$ must tend to ∞ as $t \to \infty$.

Example 1.1 Pareto distribution

We take the survivor function

$$\bar{F}(t) = (1 + t/\alpha)^{-\gamma},$$

where $\alpha > 0$ and $\gamma > 0$, simply for illustration of the basic univariate survival functions. The corresponding density and hazard functions are

$$f(t) = \gamma\alpha^{-1}(1 + t/\alpha)^{-\gamma-1}, \; h(t) = \gamma/(\alpha + t).$$

Some plots of these functions are given in Figure 1.1.

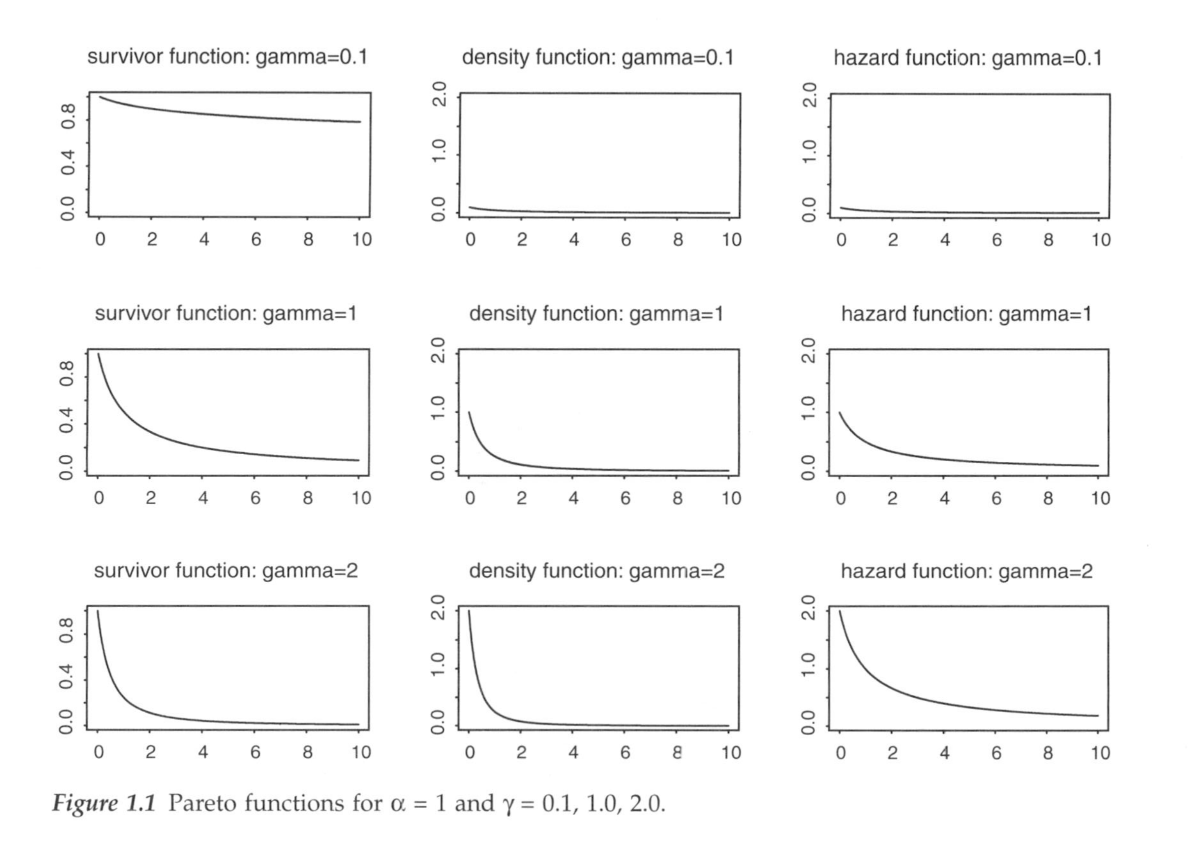

Figure 1.1 Pareto functions for $\alpha = 1$ and $\gamma = 0.1, 1.0, 2.0$.

One instance in which the Pareto distribution arises naturally in the survival context is as follows. Suppose that one begins with that old warhorse, the exponential distribution for T: $\bar{F}(t) = e^{-\lambda t}$. Suppose further that, due to circumstances beyond our control (as they say whenever I go by train), λ varies randomly over the individual systems, say with a gamma distribution to make things tractable. Thus, we have $\bar{F}(t \mid \lambda) = e^{-\lambda t}$ where λ has density

$$f(\lambda) = \Gamma(\gamma)^{-1} \alpha^{\gamma} \lambda^{\gamma-1} e^{-\alpha\lambda}.$$

Then the unconditional survivor function of T is obtained as

$$\bar{F}(t) = \int_0^{\infty} \bar{F}(t|\lambda) f(\lambda) d\lambda = \int_0^{\infty} e^{-\lambda t} \Gamma(\gamma)^{-1} \alpha^{\gamma} \lambda^{\gamma-1} e^{-\alpha\lambda} d\lambda = (1 + t/\alpha)^{-\gamma}. \blacksquare$$

The behaviour of the hazard function in Example 1.1 is fairly atypical but not altogether unknown in practice. The hazard function is decreasing and tends to zero as $t \to \infty$. Thus, it represents a system with decreasing probability of imminent failure, i.e., one that settles in as time goes on. However, it will certainly fail at some finite time since $\bar{F}(\infty) = 0$. More usually, hazard functions increase over time as systems age and become more prone to crack-up.

1.1.2 *A time and a cause: Competing Risks*

In classical Competing Risks, the observed outcome comprises T, the time to failure, and C, the cause, type, or mode of failure. The failure time T is taken to be a continuous variate here, and the cause C can take one of a fixed (small) number of values labelled 1, ..., p. The basic probability framework is thus a bivariate distribution in which one component, C, is discrete and the other, T, is continuous. It will be assumed for now that to every failure can be assigned one and only one cause from the given set of p causes. To pick a nit, they're called risks before failure and causes afterwards: the risks compete to be the cause.

For example, in Medicine, C might be the cause of death and T the age at death. In Reliability, C might identify the faulty component in a system and T the running time from start-up to breakdown. In Engineering, our subject is more likely to be referred to as the reliability of non-repairable series systems. In Economics, T might be the time spent on the unemployment register and C the reason for de-registering. In Manufacturing, T might be the usage (e.g., mileage) and C the cause of breakdown of a machine (e.g., vehicle).

The identifiable probabilistic aspect of the model is the joint distribution of C and T. This can be specified in terms of the so-called **sub-distribution functions** $F(j, t) = \text{pr}(C = j, T \le t)$, or equivalently by the **sub-survivor functions** $\bar{F}(j, t) = \text{pr}(C = j, T > t)$. These functions are related by $F(j, t) + \bar{F}(j, t) = p_j$, where

$$p_j = \text{pr}(C = j) = F(j, \infty) = \bar{F}(j, 0)$$

gives the marginal distribution of C. Thus, $F(j, t)$ is not a proper distribution function since it only reaches the value p_j instead of 1 at $t = \infty$. We assume implicitly that $p_j > 0$ and $\Sigma p_j = 1$. Note that $\bar{F}(j, t)$ is not, in general, equal to the probability that $T > t$ for failures of type j; that probability is a conditional one, $\text{pr}(T > t \mid C = j)$, expressible as $\bar{F}(j, t)/p_j$. The **sub-density function** $f(j, t)$ for continuous T is $-d\bar{F}(j, t)/dt$.

The marginal survivor function and marginal density of T can be calculated from

$$\bar{F}(t) = \sum_{j=1}^{p} \bar{F}(j, t) \quad \text{and} \quad f(t) = -d\bar{F}(t)/dt = \sum_{j=1}^{p} f(j, t).$$

The related conditional probabilities are $\text{pr}(\text{time } t \mid \text{cause } j) = f(j, t)/p_j$ and $\text{pr}(\text{cause } j \mid \text{time } t) = f(j, t)/f(t)$. These all have their uses in different contexts. For instance, $f(j, t)/p_j$ gives the distribution of age at death from cause j, which is of vital interest in medical applications (and of growing interest to me as I get older). Likewise, for a system of age t, the probability of ultimate failure from cause j is

$$\text{pr}(C = j \mid T > t) = \bar{F}(j, t)/\bar{F}(t).$$

Example 1.2 An exponential mixture

Consider the simple form $\pi_j e^{-\lambda_j t}$ for $\bar{F}(j, t)$ ($j = 1, \ldots, p$), where $\pi_1 + \ldots + \pi_p = 1$. The marginal survivor function of T is $\bar{F}(t) = \sum_{j=1}^{p} \pi_j e^{-\lambda_j t}$ and the marginal probabilities p_j for C are the π_j. The conditional distributions are defined by

$$\text{pr}(T > t \mid \text{cause } j) = e^{-\lambda_j t} \text{ and } \text{pr}(\text{cause } j \mid \text{time } t) = \pi_j \lambda_j e^{-\lambda_j t} / \sum_{k=1}^{p} \pi_k \lambda_k e^{-\lambda_k t}.$$

The model can thus be interpreted as a mixture of exponentials: the mixture proportions are the π_j, and, for a unit in the jth group, the failure time distribution is exponential with parameter λ_j. Note that C and T are not independent unless the λ_j are all equal. By differentiation, the sub-densities are $f(j, t) = \pi_j \lambda_j e^{-\lambda_j t}$.

To think about how such a model might be applied in practice, consider the case $p = 2$. Suppose that 20 failures of type 1 were recorded, 10 of type 2 were recorded, and 15 failure times were **right-censored** at fixed time t_0, i.e., all that is known about these 15 times is that they exceeded t_0. On this information alone we might be tempted to estimate π_1 as 20/30 and π_2 as 10/30. Later, after watching *Match of the Day,* it might occur to us that one team seemed to take much longer to get going than the other. So the 15 censored times might all have turned out to be of Type 2 had the game gone to extra time. In this case, we would estimate π_1 as 20/45 and π_2 as 25/45, quite different than the previous estimates. The fact that there are twice as many failures of Type 1 than of Type 2 within time t_0 has as much to do with the relative sizes of λ_1 and λ_2 as it does with those of π_1 and π_2. To make any sense out of the data, we should really try to take account of both these effects. This is just what the likelihood function does, anticipating Section 2.1. The relevant probabilities are

$$\mathrm{pr}(T > t_0) = \bar{F}(t_0) = \sum_{j=1}^{2} \pi_j e^{-\lambda_j t_0}$$

for the n_0 right-censored observations, and

$$\mathrm{pr}(C = j | T \leq t_0) = F(j, t_0)/F(t_0) = \pi_j(1 - e^{-\lambda_j t_0})/\{1 - \sum_{j=1}^{2} \pi_k e^{-\lambda_k t_0}\}$$

for the n_j observed failures of type j. Estimates based on the likelihood function

$$\{\bar{F}(t_0)\}^{n_0}\{F(1, t_0)/F(t_0)\}^{n_1}\{F(2, t_0)/F(t_0)\}^{n_2}$$

will make provision for all three independent parameters (π_1, λ_1, λ_2). In practice, we would not use this likelihood function, based only on (n_0, n_1, n_2): we would use a fuller one based on the observed times also, but the point about disentangling the parameters holds just the same. ■

1.2 *Some small data sets*

Many examples of Competing Risks analyses can be found in the mainstream statistical literature. Here, and in later chapters, a few such applications will be described in just enough detail to convey the essentials, i.e., the type of data and the methodology employed. More detail can be found in the references cited.

Data set 1 King (1971)

Table 1.1 gives breaking strengths (mg) of 23 wire connections transcribed from Nelson's (1982) Table 5.1. There are two types of failure: breakage at

Table 1.1 Breaking Strengths of Wire Connections

Bond:	0	0	550	950	1150	1150	1250	1250	1450	1450	1550	2050	3150
Wire:	750	950	1150	1150	1150	1350	1450	1550	1550	1850			

Source: From King, J. R. (1971). *Probability Charts for Decision Making*. Industrial Press, New York. With permission.

the bonded end and breakage along the wire itself. Here, the response variable T is breaking strength, or loading, rather than failure time but, as far as the statistical set-up is concerned, this is merely a change of name.

The measurements are apparently rounded to the nearest 50 mg, and the two zeros must be faulty bonds, as pointed out by Nelson who also expressed some doubts about the value 3150. ■

Data set 2 Nelson (1970)

A small subset of Nelson's data (Group 2, automatic testing) is given in Table 1.2, following Lawless' (1982, Table 10.1.2) selection. Right-censored lifetimes are indicated here, and elsewhere in this book, by listing the failure cause code as 0 (after Nelson).

Table 1.2 Failure Times and Causes (c, t) for 36 Electrical Appliances

11	1	35	15	49	15	170	6	329	6	381	6	708	6	958	10
1062	5	1167	9	1594	2	1925	9	1990	9	2223	9	2327	6	2400	9
2451	5	2471	9	2551	9	2565	0	2568	9	2694	9	2702	10	2761	6
2831	2	3034	9	3059	6	3112	9	3214	9	3478	9	3504	9	4329	9
6367	0	6976	9	7846	9	13403	0								

Source: From Nelson, W. B. (1970). *J. Qual. Technol.*, **2**, 126–149.

Nelson's main focus was to demonstrate hazard plotting techniques for data of Competing Risks type and to show how inferences may be drawn from such plots. ■

Data set 3 Nelson (1990, Problem 7.6)

Industrial heaters were subjected to temperature-accelerated life-testing with the results shown in Table 1.3. There are two failure modes, 1 denotes "open" and 2 denotes "short," and failure times are recorded in hours.

Table 1.3 Data from Accelerated Life Tests of Industrial Heaters

Temperature	Causes and Failure Times (c, t)							
1820°F	2	72.7	2	343.9	2	347.6		
1750°F	2	320.0	2	1035.0	2	1154.4	2	1979.0
1675°F	1	1532.0	1	2125.0	0	2212.0	0	2242.0
1600°F	1	1547.0	1	1726.5	2	1729.3	0	2539.0

Source: From Nelson, W. B. (1990). *Accelerated Testing: Statistical Models, Test Plans and Data Analysis*. John Wiley & Sons, New York.

Nelson set the problem of fitting an Arrhenius-log-normal model to these data and examining the goodness of fit in various ways; this will be pursued below in Section 2.4. ■

Data set 4 Hoel (1972)

Hoel's data, reproduced in Table 1.4, arise from a laboratory experiment in which male mice of strain RFM were given radiation dose 300 rads at 5 to 6 weeks old. There are two groups of mice: conventional lab environment (group 1) and germ-free environment (group 2). The survival times are measured in days and the causes of death are 1 (thymic lymphoma), 2 (reticulum cell sarcoma), and 3 (other).

Table 1.4 Survival Times of Mice

	Group 1 (n = 99)
Cause 1:	159 189 191 198 200 207 220 235 245 250 256 261 265 266 280
	343 356 383 403 414 428 432
Cause 2:	317 318 399 495 525 536 549 552 554 557 558 571 586 594 596
	605 612 621 628 631 636 643 647 648 649 661 663 666 670 695
	697 700 705 712 713 738 748 753
Cause 3:	40 42 51 62 163 179 206 222 228 252 259 282 324 333 341
	366 385 407 420 431 441 461 462 482 517 517 524 564 567 586
	619 620 621 622 647 651 686 761 763
	Group 2 (n = 82)
Cause 1:	158 192 193 194 195 202 212 215 229 230 237 240 244 247 259
	300 301 321 337 415 434 444 485 496 529 537 624 707 800
Cause 2:	430 590 606 638 655 679 691 693 696 747 752 760 778 821 986
Cause 3:	136 246 255 376 421 565 616 617 652 655 658 660 662 675 681
	734 736 737 757 769 777 800 806 825 855 857 864 868 870 870
	873 882 895 910 934 942 1015 1019

Source: From Hoel, D. G. (1972). *Biometrics*, **28**, 475–488. With permission.

Hoel developed a model that incorporates individual immunity and fitted it to the data by maximum likelihood (Section 3.5). ■

Data set 5 Boag (1949)

The survival times (months) of 121 breast cancer patients are given in Table 1.5. They come from the clinical records of one hospital over the years 1929 to 1938 and appear in Boag's Table IX under "Analysis made at February 28th, 1948." The causes of death are listed as 1 (cancer) and 2 (other). The times up to 30 months are given to the nearest tenth and after that to the nearest month.

Boag compared the fits of the log-normal distribution with others, e.g., the exponential, to the survival times. He then went on to fit the data, by maximum likelihood, with a model containing three parameters: μ and σ, for the log-normal distribution, and c, for the probability of permanent cure. Much of the paper is taken up with the problems of numerical solution of the likelihood equations in that pre-computer age. ■

Table 1.5 Survival Times of Cancer Patients

Cause 1:	0.3	5.0	5.6	6.2	6.3	6.6	6.8	7.5	8.4	8.4
	10.3	11.0	11.8	12.2	12.3	13.5	14.4	14.4	14.8	15.7
	16.2	16.3	16.5	16.8	17.2	17.3	17.5	17.9	19.8	20.4
	20.9	21.0	21.0	21.1	23.0	23.6	24.0	24.0	27.9	28.2
	29.1	30.0	31	31	32	35	35	38	39	40
	40	41	41	42	44	46	48	48	51	51
	52	54	56	60	78	78	80	84	87	89
	90	97	98	100	114	126	131	174		
Cause 2:	0.3	4.0	7.4	15.5	23.4	46	46	51	65	68
	83	88	96	110	111	112	132	162		
Cause 0:	111	112	113	114	114	117	121	123	129	131
	133	134	134	136	141	143	167	177	179	189
	201	203	203	213	228					

Source: From Boag, J. W. (1949). *J. R. Statist. Soc.*, **B11**, 15–44. © 1949 Blackwell Publishers. Reprinted with permission.

Data set 6 Nelson (1982)

The data in Table 1.6 have been taken from Problem 5.5 of Nelson (1982). They arise from a "low-cycle fatigue test of 384 specimens of a superalloy at a particular strain range and high temperature." The numbers of cycles to failure are grouped on a $\log_{10}$-scale into intervals of width 0.05, and the lower end-point of each interval is shown in the table. Failure is defined by the occurrence of a surface defect (type 1) or an interior defect (type 2). ■

Table 1.6 Log Cycles to Failure: Numbers of Specimens with Defects of Types 1 and 2

Lower	Defects		Lower	Defects		Lower	Defects	
end-point	1	2	end-point	1	2	end-point	1	2
3.55	0	3	4.00	0	16	4.45	7	2
3.60	0	6	4.05	3	12	4.50	4	3
3.65	0	18	4.10	3	4	4.55	2	1
3.70	1	23	4.15	2	3	4.60	1	0
3.75	3	47	4.20	5	10	4.65	1	0
3.80	1	44	4.25	2	3	4.70	0	0
3.85	0	57	4.30	3	4	4.75	1	0
3.90	1	37	4.35	7	7	4.80	0	0
3.95	0	20	4.40	12	4	4.85	1	0

Source: From Nelson, W. B. (1982). *Applied Life Data Analysis*. John Wiley & Sons, New York. © 1982 John Wiley & Sons, Inc. Reprinted with permission.

1.3 Hazard functions

1.3.1 Sub-hazards and overall hazard

Various hazard functions, which describe probabilities of imminent disaster, are associated with the Competing Risks set up. The **overall hazard rate** from all causes is defined by

$$\begin{aligned} h(t) &= \lim_{\delta \to 0} \delta^{-1} \mathrm{pr}(T \le t + \delta | T > t) \\ &= \lim_{\delta \to 0} \delta^{-1} \{\bar{F}(t) - \bar{F}(t+\delta)\} / \bar{F}(t) \\ &= f(t)/\bar{F}(t) = -d\log \bar{F}(t)/dt. \end{aligned} \quad (1.3.1)$$

In different contexts, $h(t)$ is variously known as the instantaneous failure rate, age-specific failure rate or death rate, intensity function, and force of mortality or decrement. The hazard function for failure from cause j, in the presence of all risks 1 to p, is defined in similar fashion by

$$\begin{aligned} h(j, t) &= \lim_{\delta \to 0} \mathrm{pr}(C = j, T \le t + \delta | T > t) \\ &= \lim_{\delta \to 0} \delta^{-1} \{\bar{F}(j, t) - \bar{F}(j, t+\delta)\} / \bar{F}(t) \\ &= f(j, t)/\bar{F}(t). \end{aligned} \quad (1.3.2)$$

Thus, $h(t) = \sum_{j=1}^{p} h(j, t)$. Some authors call the $h(j, t)$ "cause-specific" hazard functions, while others have used this term for the marginal hazards in the latent failure times set-up (see Section 3.1). To avoid ambiguity, and because the $h(j, t)$ are associated with the sub-distributions defined by the $f(j, t)$, we shall follow Crowder (1994) and call them the **sub-hazard functions**.

Example 1.2 (continued)

The sub-hazard functions for this mixture of exponentials are

$$h(j, t) = \pi_j \lambda_j e^{-\lambda_j t} / \sum_{l=1}^{p} \pi_l e^{-\lambda_l t}$$

and the overall hazard is

$$h(t) = \sum_{l=1}^{p} \pi_j \lambda_j e^{-\lambda_j t} / \sum_{l=1}^{p} \pi_l e^{-\lambda_l t}.$$

Unless the λ_j are all equal, these functions are not constant over time, this in spite of the fact that the underlying distributions are exponential. ■

In applications it is often of interest to assess the consequences of changes in certain risks. In Engineering the load profile might be altered so that some components of the system are placed under increased stress. In Medicine a treatment might improve the chances of surviving a particular disease. Suppose that the sub-hazard function $h(j, t)$ is increased from its previous level over the time interval $I = (a, b)$ and that the other sub-hazards are unchanged. It is reasonable to conjecture that the overall probability of failure in I is increased and that the relative probability of failure from a cause other than j in I is decreased. This is proved to be true in the following lemma which serves as an example of this type of calculation. The opposite conclusions would hold if $h(j, t)$ were decreased over I, of course.

Lemma 1.1 (Kimball, 1969)

Suppose that $\int_a^b h(c, t)dt$ is increased for $c = j$ only. Then (i) P_I, the probability of failure in $I = (a, b)$, conditional on survival to enter I, is increased, and (ii) P_{Ik}, the probability of failure in I from cause k, conditional on entry to I, is decreased for $k \neq j$.

Proof

We have

$$P_I = \text{pr}(T \le b | T > a) = \{\bar{F}(a) - \bar{F}(b)\}/\bar{F}(a) = 1 - \exp\left\{-\int_a^b h(t)dt\right\},$$

using Equation 1.1.1. Since $h(t) = \sum_j h(j, t)$, the specified increase results in a decrease in the exponential term, and so (i) is verified. For (ii), we have

$$\begin{aligned} P_{Ik} &= \text{pr}(C = k, T \le b | T > a) = \{\bar{F}(k, a) - \bar{F}(k, b)\}/\bar{F}(a) \\ &= \int_a^b f(k, t)dt/\bar{F}(a) - \int_a^b h(k, t)\{\bar{F}(t)/\bar{F}(a)\}dt. \end{aligned}$$

Now, $h(k, t)$ is unchanged from its previous level, whereas $\bar{F}(t)/\bar{F}(a)$ is increased, by the same argument as used for (i). Hence the result. ■

Stochastic models for Competing Risks can be specified directly in terms of the sub-hazards, rather than the sub-survivor functions or sub-densities, as in the following example.

Example 1.3 Weibull sub-hazards

Suppose that the form $\phi_j \lambda_j^{\phi_j} t^{\phi_j - 1}$ is adopted for $h(j, t)$. (This form is just used here as an illustration. A discussion of ways of choosing suitable hazard forms will be given later in Chapter 6.) Then the overall hazard function is $h(t) = \sum_{j=1}^{p} \phi_j \lambda_j^{\phi_j} t^{\phi_j - 1}$. The marginal survivor function of T follows from Equation 1.3.1 as

$$\bar{F}(t) = \exp\left\{-\int_0^t h(s)ds\right\} = \exp\left\{-\sum_{j=1}^{p} (\lambda_j t)^{\phi_j}\right\} = \prod_{j=1}^{p} \exp\{-(\lambda_j t)^{\phi_j}\},$$

which shows that T has the distribution of $\min\{T_1, \ldots, T_p\}$, the T_j being independent Weibull variates. The sub-densities $f(j, t)$ can be calculated as $h(j, t)\bar{F}(t)$, but integration of this to find $\bar{F}(j, t)$ is generally intractable.

Note that this model is not the same as the one obtained by extending the exponential mixture of Example 1.2 to its Weibull counterpart, which has sub-survivor functions of form $\pi_j \exp\{-(\lambda_j t)^{\phi_j}\}$ for $j = 1, \ldots, p$. ■

1.3.2 *Proportional hazards*

The relative risk of failure from cause j at time t is $h(j, t)/h(t)$. If this is independent of t for each j, then **proportional hazards** are said to obtain. In this case, as time marches on, the relative risks of the various causes of failure stay in step, no one increasing its share of the overall risk, though the latter might rise or fall. This is beginning to sound a bit like independence of cause and time of failure, and it is proved to be so in Theorem 1.1 below.

In many areas of application, proportional hazards would probably be the exception rather than the rule. For instance, in the health sciences, the relative risks of cot death and senile dementia might be expected to vary with age. The assumption, at least in piecewise fashion along the time scale, has been attributed to Chiang (1961) by David (1970). Seal (1977, Section 3) indicated that it goes back much further.

Example 1.3 (continued)

Proportional hazards obtains only if the ϕ_j are all equal, say to ϕ. In this case,

$$\bar{F}(t) = \exp\left\{-(\sum_{j=1}^{p} \lambda_j^{\phi})t^{\phi}\right\} \quad \text{and} \quad \bar{F}(j, t) = \pi_j \exp\left\{-(\sum_{j=1}^{p} \lambda_j^{\phi})t^{\phi}\right\},$$

where $\pi_j = \lambda_j^{\phi} / \sum_{k=1}^{p} \lambda_k^{\phi}$. This is a mixture model in which $p_j = \text{pr}(C = j) = \pi_j$ and

$$\text{pr}(T > t | C = j) = \text{pr}(T > t) = \exp\left\{-(\sum_{j=1}^{p} \lambda_j^{\phi})t^{\phi}\right\}. \blacksquare$$

Theorem 1.1 (Elandt-Johnson, 1976; David and Moeschberger, 1978; Kochar and Proschan, 1991)

The following conditions are equivalent:

(i) proportional hazards obtains;
(ii) the time and cause of failure are independent;
(iii) $h(j, t)/h(k, t)$ is independent of t for all j and k.

In this case $h(j, t) = p_j h(t)$ or, equivalently, $f(j, t) = p_j f(t)$, or $F(j, t) = p_j F(t)$.

Proof

For (i) and (ii), we have

$$\text{pr}(\text{cause } j \mid \text{time } t) = f(j, t)/f(t) = h(j, t)/h(t);$$

independence of t must obtain for both sides of this equation or for neither. The last statement of the theorem follows since the left hand side equals p_j in this case. For (i) and (iii), note that

$$h(j, t)/h(t) = h(j, t)/\sum_{k=1}^{p} h(k, t)$$

is independent of t for all j if and only if (iii) holds. ■

Part (ii) of the theorem says, for instance, that failure during some particular period does not make it any more or less likely to be from cause j than failure in some other period.

Another hazard function that could be defined is

$$g(j, t) = f(j, t)/\bar{F}(j, t) = -d\log\bar{F}(j, t)/dt.$$

This looks more like the usual form (Section 1.1.1); it is the hazard for failures from cause j only. However, as will be seen (e.g., Theorems 3.3 and 7.1), it is the $h(j, t)$ that arise naturally in the development of the theory. Also, the conditioning event $\{T > t\}$ used in $\text{pr}(C = j, T \le t + \delta \mid T > t)$ to define $h(j, t)$ is more relevant in practice than the conditioning event $\{C = j, T > t\}$ used above: at time t, you know that failure has not yet occurred, but you probably do not know what the eventual cause of failure will turn out to be.

1.4 Regression models

In most applications, the data are not independently, identically distributed and a regression model is called for to relate the basic parameters to covariates or explanatory variables. Suppose that to the ith case in the sample is attached a vector x_i of such explanatory variables. Thus, x_i might contain components recording the age, height, weight, and health history of a patient or the age, dimensions, specification, and usage history of an engineering system component.

Example 1.2 (continued)

A regression model for the exponential mixture might be specified in terms of log-linear forms as (i) $\log\lambda_{ij} = x_i^T\beta_j$, and (ii) $\log\pi_{ij} = \kappa_i + x_i^T\gamma_j$. In (ii) κ_i is determined by the requirement that $\pi_{i+} = \pi_{i1} + \ldots + \pi_{ip} = 1$ as

$$\kappa_i = -\log\sum_{j=1}^{p} \exp(x_i^T\gamma_j),$$

so the model for π_{ij} is expressible as

$$\pi_{ij} = \exp(x_i^T\gamma_j)/\sum_{l=1}^{p} \exp(x_i^T\gamma_l);$$

an identifiability constraint, such as $\gamma_p = \mathbf{0}$, is required.

If x_i is of dimension q, the full parameter set, comprising vectors $\beta_1, \ldots, \beta_p$ and $\gamma_1, \ldots, \gamma_{p-1}$, will be of dimension $(2p - 1)q$. This will often be too large for the limited set of data at hand. In that case, further constraints will be necessary to reduce the number of parameters to be estimated. A typical reduction, applicable in the usual case where x_i has first component 1, is to take $\beta_j = (\beta_{0j}, \beta)$; here, β_{0j} is a scalar "intercept" parameter, allowed to vary over j, and β is a vector of regression coefficients of length $q - 1$, the same for all j. A similar reduction applied to the γ_j is not sensible: it would effectively remove all the γ-regression coefficients, since they would be "divided out" in the expression above for π_{ij}, just leaving the intercepts γ_{0j}. An alternative reduction is to set selected components of the β_j and γ_j to zero. This is applicable when the corresponding components of x_i are known to have no effect on the associated λ or π. ■

Example 1.4 Larson and Dinse (1985)

These authors used a mixture model like that of Example 1.2. They justified it as being appropriate for situations in which the eventual failure cause is determined at time zero by some random choice. An example that they quote is where, following treatment for a terminal disease, a patient might be cured or not and this can only be determined at some future date. Their model, which allows for a vector x of explanatory variables, is as follows:

(i) $\text{pr}(C = j; x) = p_{jx} = \phi_{jx}/\sum_{l=1}^{p}\phi_{lx}$, where $\phi_{jx} = \exp(x^T\beta_j)$ and $\beta_p = 0$ for identifiability;

(ii) $\text{pr}(T > t \mid C = j; x) = \exp\{-\psi_{jx}G_j(t)\}$, where $G_j(t) = \int_0^t g_j(s)ds$, $g_j(s)$ is piecewise constant in s and $\psi_{jx} = \exp(x^T\gamma_j)$.

Thus, loglinear models are used for ϕ_{jx} and ψ_{jx}, with regression coefficients β_j and γ_j, respectively. From these specifications follow

$$\bar{F}(j, t) = p_{jx}\exp\{-\psi_{jx}G_j(t)\},\ f(j, t) = p_{jx}\psi_{jx}g_j(t)\exp\{-\psi_{jx}G_j(t)\},$$
$$\bar{F}(t) = \sum\nolimits_{j=1}^{p} p_{jx}\exp\{-\psi_{jx}G_j(t)\},$$

and hence $h(j, t)$, etc.

Larson and Dinse computed maximum likelihood estimates via an EM algorithm which replaces right-censored failure times with unknown cause by a set of suitably weighted hypothetical ones with known cause. They applied their analysis to some heart transplant data with two causes of death: transplant rejection and other causes. There were $n_1 = 29$ rejection cases, $n_2 = 12$ deaths from other causes, and $n_0 = 24$ patients still alive at the end of the study. The explanatory variables comprised a mismatch score (comparing donor-recipient tissue compatibility), age, and waiting time (from admission to the program to transplantation). They also compared their parametric model-based analysis of these data with a semi-parametric analysis. ■

We will now consider some general classes of regression models for Competing Risks. These are adaptations of ones which have been widely used in univariate Survival Analysis. For generally dependent risks it is the sub-distributions that determine the structure, so it is in terms of these that the models are framed. Further discussion of these classes of models for the univariate case, including details of diagnostic plots, is given in Crowder et al. (1990, Chapter 4).

1.4.1 Proportional hazards

The univariate version is $h(t;\boldsymbol{x}) = \psi_x h_0(t)$, where h_0 is some baseline hazard function and ψ_x is a positive function of the vector $\boldsymbol{x}$ of explanatory variables, e.g., $\psi_x = \exp(\boldsymbol{x}^{\mathrm{T}}\beta)$. Note that "proportional hazards" has a different meaning in univariate models to that given in Section 1.3 for Competing Risks. A natural adaptation of the univariate definition to the case of Competing Risks can be made in terms of the sub-hazard functions:

$$h(j, t;\boldsymbol{x}) = \psi_{jx}h_0(j, t).$$

A second stage can be imposed, taking "proportional hazards" now in the Competing Risks sense, to replace $h_0(j, t)$ by $p_j h_0(t)$:

$$h(j, t;\boldsymbol{x}) = \psi_{jx}p_j h_0(t).$$

In a statistical analysis, with parametrically specified h-functions, one can test for these successive restrictions on the sub-hazard models in the usual parametric way.

1.4.2 Accelerated life

This model specifies the form $\bar{F}(t;\boldsymbol{x}) = \bar{F}_0(\psi_x t)$ for a univariate survivor function, with $\boldsymbol{x}$ and ψ_x as for proportional hazards and $\bar{F}_0$ some baseline survivor function. The effect of $\boldsymbol{x}$ is to accelerate the time scale t by a factor ψ_x in the baseline model. A natural analogue for Competing Risks is

$$\bar{F}(j, t;\boldsymbol{x}) = \bar{F}_0(j, \psi_{jx} t)$$

in terms of baseline sub-survivor functions $\bar{F}_0(j, t)$. A second, proportional hazards, stage can be incorporated as in Section 1.4.1 to yield

$$\bar{F}(j, t;\boldsymbol{x}) = \bar{F}_{00}(\psi_{jx} t)^{pj}$$

for some $\bar{F}_{00}$.

The regression version of Example 1.2 given above is an accelerated life model provided that $\pi_{ij} = \pi_j$ for all i; in this case $\bar{F}_0(j, t)$ can be taken as $\pi_j e^{-t}$ with $\psi_{jx} = \exp(\boldsymbol{x}_i^T \beta_j)$.

1.4.3 Proportional odds

In the univariate version of this model the odds on the event $\{T \le t\}$ are expressed as

$$\{1 - \bar{F}(t;\boldsymbol{x})\}/\bar{F}(t;\boldsymbol{x}) = \psi_{jx}\{1 - \bar{F}_0(t)\}/\bar{F}_0(t).$$

A natural Competing Risks version of this is

$$\{1 - \bar{F}(j, t;\boldsymbol{x})\}/\bar{F}(j, t;\boldsymbol{x}) = \psi_{jx}\{1 - \bar{F}_0(j, t)\}/\bar{F}_0(j, t).$$

A second, proportional hazards, stage can be incorporated as in Section 1.4.1, replacing $\bar{F}_0(j, t)$ here by $\bar{F}_{00}(t)^{p_j}$.

1.4.4 Mean residual life

The **mean residual lifetime** at age t is defined as

$$m(t) = E(T - t \mid T > t);$$

this is the (expected) time left to an individual who has survived to age t; the corresponding **life expectancy** is $E(T \mid T > t) = m(t) + t$. The mrl can be evaluated as

$$m(t) = \int_t^{\infty}(y-t)f(y|y>t)dy = \int_t^{\infty}(y-t)\{f(y)/\bar{F}(t)\}dy$$
$$= \{\bar{F}(t)\}^{-1}\int_t^{\infty}\bar{F}(y)dy \tag{1.4.1}$$

where $f(y \mid y > t)$ is the indicated conditional density of T and an integration-by-parts has been performed under the assumption that $y\bar{F}(y) \to 0$ as $y \to \infty$. Obviously, $m(0) = E(T)$ and, differentiating Equation 1.4.1 with respect to t,

$$m'(t)\bar{F}(t) + m(t)\bar{F}'(t) = -\bar{F}(t),$$

which yields

$$\{m'(t)+1\}/m(t) = -\bar{F}'(t)/\bar{F}(t) = h(t),$$

and thence, by integration,

$$\log\{m(t)/m(0)\} + \int_0^t m(y)^{-1}dy = -\log\{\bar{F}(t)\} = H(t);$$

note that a constant of integration has been included to match the two sides at $t = 0$. Hence, we get the inverse relationship to Equation 1.4.1 as

$$\bar{F}(t) = \{m(t)/m(0)\}^{-1}\exp\left\{-\int_0^t m(y)^{-1}dy\right\}. \tag{1.4.2}$$

For regression, the mean residual lifetime function can be modelled as follows, following Oakes and Dasu (1990) and Maguluri and Zhang (1994):

$$m(t;x) = \psi_x m_0(t),$$

where m_0 is some baseline mean residual life function and ψ_x is a positive function of the vector x of explanatory variables, e.g., $\psi_x = \exp(x^T\beta)$.

The above is the univariate version. To extend this to Competing Risks, we can define sub-mrl functions as

$$m(j,t) = \int_t^{\infty}(y-t)f(j,y|y>t)dy = \{\bar{F}(t)\}^{-1}\int_t^{\infty}\bar{F}(j,y)dy; \tag{1.4.3}$$

then, $\sum_{j=1}^{p} m(j,t) = m(t)$. For the inverse relationship, giving $\bar{F}(j, t)$ in terms of the $m(j, t)$, we first derive

$$m'(j,t)\bar{F}(t) + m(j,t)\bar{F}'(t) = -\bar{F}(j,t),$$

by differentiation of Equation 1.4.3. Then, using the relation

$$\bar{F}'(t) = -\bar{F}(t)\{m'(t) + 1\}/m(t)$$

from above, obtain

$$\bar{F}(j,t)/\bar{F}(t) = m(t)^{-1}\{m'(t) + 1\}m(j,t) - m'(j,t).$$

The proportional mrl's model can be extended straightforwardly to Competing Risks as

$$m(j,t;\boldsymbol{x}) = \psi_{jx} m_0(j,t).$$

However, as with all the models described here, without a real application with a convincing interpretation for the proposed form, one has only a theoretical construction.

chapter 2

Parametric likelihood inference

We will only use likelihood-based methods in this book. By this is meant both Frequentist, relying on the usual asymptotic approximations, and Bayesian, relying on appropriate priors. There are many useful ad hoc methods for obtaining answers to particular questions, but we will stick to the general approach, which is capable of giving answers to any well-formulated parametric question. However valid the methodology, of course, the data need to be sufficiently informative and the model has to be sufficiently well-fitting. How many seminars have you attended where the conclusion was "Nice method, shame about the data?"

2.1 The likelihood for Competing Risks

2.1.1 Forms of the likelihood function

Suppose that data $\{(c_i, t_i) : i = 1, \ldots, n\}$ are available and that parametric models have been specified for the sub-densities $f(j, t)$. Then the likelihood function is given by

$$L = \Pi_o f(c_i, t_i) \times \Pi_c \bar{F}(t_i) . \tag{2.1.1}$$

Here Π_o denotes the product over cases for which failure has been observed and Π_c denotes the product over right-censored failure times of cases where failure has not yet been observed. The latter can conveniently be indicated in the data by $c_i = 0$, and t_i is then recorded as the time at which observation ceased. It is sometimes not appreciated that events that have not yet occurred, like unobserved failures, can provide information. Sherlock Holmes was well aware of this: he gained a vital clue from the "curious incident" that the dog did not bark in *Silver Blaze* (Doyle, 1950). The censoring mechanism is assumed here to operate independently of the failure mechanisms: a fuller discussion of censoring is given in Section 7.6.

The likelihood can be expressed in terms of the sub-hazard functions: using $f(j, t) = h(j, t)\bar{F}(t)$, Equation 2.1.1 can be written as

$$L = \Pi_o h(c_i, t_i) \times \Pi_{i=1}^n \bar{F}(t_i)\,. \tag{2.1.2}$$

Further, in the second term here, $\bar{F}(t_i)$ can be written as $\exp\{-\int_o^t h(s)ds\}$ with $h(s) = h(1, s) + \ldots + h(p, s)$. Thus, an alternative specification for the model can be made entirely in terms of the $h(j, t)$ rather than the $f(j, t)$. This approach has advantages that will be discussed in Chapter 6.

Another way of writing the likelihood function is

$$L = \Pi_{i=1}^n \{h(c_i, t_i)^{d_i} \bar{F}(t_i)\} \tag{2.1.3}$$

where d_i is a censoring indicator for case i: $d_i = 1$ if failure has been observed, $d_i = 0$ if the failure time is right-censored. Again, Equation 2.1.3 can be re-expressed as $L = L_1 L_2 \ldots L_p$ with

$$L_j = \Pi_j h(j, t_i) \times \Pi_{i=1}^n \exp\left\{-\int_o^{t_i} h(j, s)ds\right\}, \tag{2.1.4}$$

where Π_j denotes the product over $\{i : c_i = j\}$. Here, L_j accounts for the contributions involving the sub-hazard $h(j, t)$. If, as is often the case, the model has been parametrized so that the $h(j, t)$ have non-overlapping parameter sets, such a partition of L can be useful. Numerically, it might be much easier to maximize each L_j separately, since there are fewer parameters to handle at any one time. Sampford (1952) made this point in connection with his "multi-stimulus" model. Theoretically, the parameter sets will be orthogonal, which can facilitate inference considerably (Cox and Reid, 1987). Thus, for example, if we are particularly interested in one of the failure causes, say the jth, the parameters governing $h(j, t)$ can be estimated and examined without reference to those of the other failure causes. In fact, one does not even have to specify parametric forms for the other $h(k, t)$ at all. Obviously, the greater the number of observations contributing to L_j, the better the estimates of the parameters of $h(j, t)$. Observed failures due to an identified cause are more useful in this respect than right-censored failure times. From the Statistician's point of view, then, censoring is a fate worse than death.

2.1.2 *Incomplete observation of C or T*

(a) *Uncertainty about C*

In some cases it is not altogether clear after a system failure what exactly went wrong. For instance, after a machine breakdown, the fault might be narrowed down only to one of several possibilities. This is commonly the case where failure is destructive in some way, such as in a plane crash. Again, relatives might refuse to allow a post-mortem, thus leaving the cause of death uncertain. In such cases, and provided that the observed time and set of possible causes gives no additional information about which individual cause was the culprit, the likelihood contribution for an observed failure is just modified from $f(j, t)$ to $\Sigma f(j, t)$, with summation over the set of possible indices j. Clearly, it is likely that inference will suffer as the amount of uncertainty in the data increases. In the extreme, where the failure cause is not documented at all, the likelihood involves only the function $\bar{F}(t)$ and its derivative, i.e., inference is only possible about the marginal distribution of T. In this case we are back to standard Survival Analysis.

(b) Uncertainty about T

The instant of failure can be missed because the watchman was asleep or the recording equipment was on the blink. A system might only be inspected periodically, a breakdown in the meantime only causing a minor, unnoticed holocaust. All that is known is that the failure happened between times t_1 and t_2, say. The likelihood contribution for a failure from cause j during the time interval (t_1, t_2) is $\bar{F}(j, t_1) - \bar{F}(j, t_2)$. It can even happen that the failure cause is known ahead of the failure, the time being right-censored (e.g., Dinse, 1982); then t_2 is taken as ∞. If there is also uncertainty about the cause of failure, this difference must be summed over the possible indices j.

2.1.3 Maximum likelihood estimates

We consider first a very simple, and therefore atypical, example just to illustrate the derivation of maximum likelihood estimators.

Example 2.1

Recall the model of Example 1.3, involving Weibull sub-hazards with $h(j, t) = \phi_j \lambda_j^{\phi_j} t_j^{\phi_j - 1}$ and $\bar{F}(t) = \exp\{-\sum_{j=1}^{p} (\lambda_j t)^{\phi_j}\}$. From Equation 2.1.4, the log-likelihood contributions take the form

$$\log L_j = n_j \log \phi_j + n_j \phi_j \log \lambda_j + (\phi_j - 1) \sum_j \log t_i - \lambda_j^{\phi_j} s_j,$$

where n_j is the number of observed failures from cause j and $s_j = t_1^{\phi_j} + \ldots + t_n^{\phi_j}$. Provided that λ_j does not appear as a parameter in any other $h(k, t)$, we have

$$d\log L / d\lambda_j = d\log L_j / d\lambda_j = n_j \phi_j / \lambda_j - s_j,$$

and so the maximum likelihood estimate $\hat{\lambda}_j$ for λ_j can be written down explicitly in terms of ϕ_j as $n_j \phi_j / s_t$. That for ϕ_j requires numerical maximization of $\log L$. In the particular exponential case, $\phi_j = 1$, $\hat{\lambda}_j = n_j / t_+$, where

$t_+ = t_1 + \ldots + t_n$ is the total time logged, or "total time on test," including censored times. ■

Simplifications, such as the ability to solve likelihood equations explicitly for some parameters, are not typically possible for Competing Risks. This is illustrated by the complicated likelihood function which results from a simple model in the following example.

Example 2.2

The parameter set of the regression model given for Example 1.2 in Section 1.4 comprises $(\beta_1, \ldots, \beta_p)$ and $(\gamma_1, \ldots, \gamma_p)$, and the corresponding log-likelihood function is

$$\begin{aligned}\log L &= \Sigma_o \log(\pi_{ic_i}\lambda_{ic_i}e^{-\lambda_{ic_i}t_i}) + \Sigma_c \log(\Pi_{k=1}^p \pi_{ik}e^{-\lambda_{ik}t_i}) \\ &= \Sigma_o\left\{x_i^T\gamma_{c_i} - \log\kappa_i + x_i^T\beta_{c_i} - t_i e^{-x_i^T\beta_{c_i}}\right\} \\ &\quad + \Sigma_c \log\left\{\Pi_{k=1}^p \kappa_i^{-1}\exp(x_i^T\gamma_k - t_i e^{-x_i^T\beta_k})\right\}.\end{aligned}$$

The derivatives of $\log L$ with respect to the parameters are rather unwieldy algebraic expressions and explicit formulae for the maximum likelihood estimators are not obtained. ■

There is plenty of scope for getting knee-deep in algebra when investigating particular models in the likelihood function, e.g., see David and Moeschberger (1978). Without mentioning any names, some of the literature puts one in mind of the sons of toil shifting tons of soil. In this book such algebraic heroics will be avoided. We will take the view that, in computing terms, one just needs a subroutine or procedure that will return a value for the likelihood function on call. The rest should be left to the professionals, the numerical analysts and computer programmers who have developed powerful methods for function optimisation over the last quarter of a century. The general problem of computing maximum likelihood estimates numerically is discussed in more detail in Appendix 1.

2.2 Model checking

2.2.1 Goodness of fit

A well-worn approach to goodness of fit, which has worn well, is to compare the model fitted with an extended, more general one. For instance, this is what one is doing when assessing the effect of additional explanatory variables in regression. Again, one might fit Weibull sub-distributions, extending the model of Example 1.2, with $\bar{F}(j, t) = \pi_j \exp\{-(\lambda_j t)^{\phi_j}\}$, and then test to see whether they can be downgraded to exponentials, with $\phi_j = 1$. Another standard method is to employ a so-called **Lehmann alternative**: when the model is based on a particular distribution function $F(t)$, a modified model, based on $F(t)^\phi$, where ϕ is an extra parameter, can be entertained. If $F(t)$ already has a power law form, the Lehmann alternative model will be no more general than the original one; in this case, the transformation can be performed on the survivor function, $\bar{F}(t)$. The original model will be supported if, when the alternative is fitted and tested, $\phi = 1$ is acceptable.

For Competing Risks, various comparisons can be made of estimated quantities with their observed counterparts, e.g., $p_j = \bar{F}(j, 0)$ (the proportion of failures from cause j), $\bar{F}(j, t)/p_j$ (the survivor function of failures from cause j), and $\bar{F}(t)$ (the survivor function for all failures). In the survivor function plots, right-censored observations might have to be accommodated by appropriate modification of the empirical survivor functions.

2.2.2 *Residuals*

Residuals can be computed for Competing Risks data by an extension of the methods used for univariate Survival Analysis. Thus, the survivor function for failures of type j is the conditional one, $p_j^{-1}\bar{F}(j, t)$; this is just $\bar{F}(t)$ in the case of proportional hazards. Let the observed failure times of type j be t_{ij} ($i = 1, \ldots, n_j$) and define $e_{ij} = p_{ij}^{-1}\bar{F}_i(j, t_{ij})$; this is just a **probability integral transform** of t_{ij}, using the survivor function instead of the distribution function. The subscript i on $\bar{F}_i$ and p_{ij} indicates that these may be functions of explanatory variables $\mathbf{x}_i$, so that the ordering among the e_{ij} might not be the same as that originally among the t_{ij}. With estimates inserted for the parameters, the $\hat{e}_{ij}$ should resemble a random sample of size n_j from U(0, 1), the uniform distribution on (0, 1). However, censoring, as usual, muddies the water. Of the n_0 censored failure times t_{i0}, some would eventually have turned out to be failures from cause j. If we knew which ones, then we could transform them to $e_{i0j} = p_{ij}^{-1}\bar{F}_i(j, t_{i0})$, and put them in with the uncensored e_{ij}, after making the adjustment of replacing e_{i0j} by $(e_{i0j} + 1)/2$; this is obtained by adding the mean residual lifetime (Section 1.4.4), $(1 - e_{i0j})/2$, to e_{i0j}. Since we do not know which ones, we could arbitrarily select the "most likely" ones, i.e., those for which q_{ij} is largest among the q_{ik} ($k = 1, \ldots, p$), where

$$q_{ij} = \text{pr}_i(C = j | T > t_{i0}) = \bar{F}_i(j, t_{i0}) / \bar{F}_i(t_{i0}).$$

In this way, we would obtain an approximation to the full set of residuals from failures of type j. More favourably, this arbitrary selection from the censored values can be avoided if the e_{i0j} ($i = 1, \ldots, n_0$) are all larger than the e_{ij}. This would certainly obtain, for instance, if there were no explanatory variables and all failure times were censored at the same value. Then the e_{ij} are the first n_j order statistics from a uniform sample of size $m_j = n_j + q_j n_0$, where q_j is the proportion of censored values which would eventually be type j failures. If we knew q_j, we could then plot residuals as the points $\{i/(m_j + 1), \hat{e}_{ij}\}$, $i/(m_j + 1)$ being the expected value of e_{ij}. What we can do is to insert an estimate of q_j, an obvious one being $q_{j1} + \ldots + q_{jn_0}$. If there is only a small amount of overlap between the e_{i0j} and the e_{ij}, we could reduce the cut-off point to the lowest e_{i0j}-value, and treat the e_{ij}s beyond as censored. Evidently, the situation is not ideal for generating residuals and, in practice, some ad hoc measures will often be needed.

Many other types of residuals and diagnostics are used in applied work. The books by Cook and Weisburg (1982) and Atkinson (1985) contain a wealth of material, and an overview is given by Davison and Snell (1990). Among the other types of residual are leave-one-out residuals, where each observation is compared with its value predicted from the model fitted to the rest (a type of cross-validation), and deviance residuals, or log-likelihood contributions to us traditionalists. Besides plotting, residuals can be used to construct test statistics. A useful one, based on uniform residuals, whose large-sample distribution is not distorted by replacing θ by $\hat{\theta}$ in F, is Moran's (Cheng and Stevens, 1989).

2.3 Inference

2.3.1 Hypothesis tests and confidence intervals

Apart from goodness of fit of the broad class of model, there are the particular effects of explanatory variables to be examined, when such are present. The first question for each is often whether the corresponding regression coefficient is zero or not. Appropriate likelihood ratio tests can be applied or asymptotic equivalents such as those based on the score function and the maximum likelihood estimators. The latter are less well recommended though, in view of their lack of invariance under parametric transformation, e.g., Cox and Hinkley, 1974, Section 9.3 (vii).

Confidence intervals for individual parameters can be obtained via the usual large-sample approximations; standard errors can be computed from the inverse of $\partial^2 \log L/\partial\theta^2$, the second derivative (Hessian) matrix of the log-likelihood with respect to the parameter set θ. As when fitting the model, the differentiation can be performed numerically by differencing (Appendix 1). The observed information matrix, rather than its expected value, is preferred here (Efron and Hinkley, 1974). The asymptotic normal approximation to the distribution of the maximum likelihood estimators, on which this is based, can sometimes be usefully improved by transformation of the parameters, e.g., Cox and Hinkley, 1974, Section 9.3 (vii).

2.3.2 Bayesian approach

A general review of Bayesian ideas and decision analysis for reliability and survival data is given in Chapter 6 of Crowder et al. (1991), and the book by Martz and Waller (1982) deals with the whole subject in much more detail and gives many references to applications. Singpurwalla (1988) makes a more fundamental case for the subjective approach.

Moving on to Competing Risks, the literature does not seem to be overburdened with Bayesian applications, though there would not appear to be any special technical problems with the set-up. However, just as the likelihood equations are usually intractable, necessitating numerical solutions, so are the Bayesian manipulations analytically impossible for all but the very simplest models. In a second "however," however, great strides have been made over the past few years in developing practical computing strategies for Bayesian analysis. Before going on to computing strategies, we give an atypically tractable example.

Example 2.3 (Bancroft and Dunsmore, 1976)

Consider the model with sub-survivor functions $\bar{F}(j, t) = p_j e^{-\lambda_+ t}$, where $p_j = \lambda_j/\lambda_+$ and $\lambda_+ = \lambda_1 + \ldots + \lambda_p$. This is a special case of Example 1.2. The likelihood function for data $\{(c_i, t_i) : i = 1, \ldots, n\}$ is, from Equation 2.1.1,

$$L(\lambda) = \Pi_{c_i \neq 0} f(c_i, t_i) \times \Pi_{c_i = 0} \bar{F}(t_i) = \Pi_{c_i \neq 0}\{\lambda_{c_i} e^{-\lambda_+ t_i}\} \times \Pi_{c_i = 0} e^{-\lambda_+ t_+}$$

$$= \{\Pi_{j=1}^{p} \lambda_j^{n_j}\} e^{-\lambda_+ t_+},$$

where $\lambda = (\lambda_1, \ldots, \lambda_p)$, $t_+ = t_1 + \ldots + t_n$ and n_j is the number of failures of type j in the sample. For a Bayesian analysis, we need a joint prior for the λ_j. A natural conjugate prior has the form of independent gamma distributions:

$$f(\lambda) \propto \Pi_{j=1}^{p}\{\lambda_j^{\gamma_j - 1} e^{-\nu_j \lambda_j}\}.$$

The joint posterior distribution for the λ_j is now given by

$$f(\lambda | \text{data}) \propto f(\lambda) L(\lambda) \propto \Pi_{j=1}^{p}\{\lambda_j^{\gamma_j + n_j - 1} e^{-(\nu_j + t_+)\lambda_j}\},$$

representing independent gamma distributions for the λ_j with modified shape and scale parameters. Bancroft and Dunsmore derived the predictive density for a future observation (C, T):

$$f(c, t | \text{data}) = \int f(c, t | \lambda, \text{data})\ f(\lambda | \text{data}) d\lambda$$

$$\propto \int \lambda_c e^{-\lambda_+ t} \Pi_{j=1}^{p}\{\lambda_j^{\gamma_j + n_j - 1} e^{-(\nu_j + t_+)\lambda_j}\} d\lambda = \Pi_{j=1}^{p}\{(t + \nu_j + t_+)^{-a_{cj}} \Gamma(a_{cj})\}$$

where $\lambda_0 = 1$ and $a_{cj} = \delta_{cj} + \gamma_j + n_j$, δ_{cj} being the Kronecker delta taking value 1 if $c = j$ and 0 otherwise. If $\nu_j = \nu$ for all j, C and T are conditionally independent, given the data, with

$$\text{pr}(C = c|\text{data}) \propto \Pi_{j=1}^{p}\Gamma(a_{cj}),$$

which leads to

$$\text{pr}(C = c \mid \text{data}) = (1 - \delta_{c0})(\gamma_c + n_c)/a_+ + \delta_{c0}/a_+,$$

and T having Pareto density

$$f(t|\text{data}) = (a_+ - 1)^{-1}(t + \nu + t_+)^{-a_+}$$

on $(0, \infty)$, where $a_+ = 1 + \gamma_+ + n_+$, $\gamma_+ = \gamma_1 + \ldots + \gamma_p$ and $n_+ = n_1 + \ldots + n_p$. ■

2.3.3 *Bayesian computation*

Until fairly recently, say 20 years ago, the Bayesian approach was severely hampered by the computational difficulties involved. However, a revolution has taken place — Markov chain Monte Carlo has arrived. Its development has been driven by a number of people, Gelfand and Smith (1990), Smith (1991), Geyer (1992) and Smith and Roberts (1993), to mention just a few. The book by Gilks et al. (1996) is a good all-round introduction: the basic approach is set out and many real applications are described in detail. Some notes are given in Appendix 2.

The output from the McMC computations is a (large) sample, say $(\theta_1, \ldots, \theta_m)$, from the joint posterior distribution of the parameters. This can be used to estimate parametric functions of interest. For instance, the posterior mean of such a function, say $\eta(\theta)$, is given by

$$\int \eta(\theta)p(\theta|D)d\theta,$$

where $p(\theta \mid D)$ denotes the posterior density of θ given the data D. This integral can be estimated by the sample mean of the values $\eta(\theta_j)$ ($j = 1, \ldots, m$), i.e., by

$$m^{-1}\sum_{j=1}^{m}\eta(\theta_j).$$

For a particular example, suppose that we are interested in the probability that θ_1/θ_2 lies between a and b. This is expressible in the integral form above by taking $\eta(\theta) = \text{I}\{a < \theta_1/\theta_2 < b\}$, where I{.} denotes the indicator function. The posterior estimate now follows as above; in this case it reduces to the proportion of sampled parameter vectors for which $a < \theta_1/\theta_2 < b$. For another example, to predict a future "independent" observation, say y, we can use its posterior density,

$$p(y|D) = \int p(y|\theta)p(\theta|D)d\theta,$$

in which it has been assumed that $p(y \mid D, \theta) = p(y \mid \theta)$. This corresponds to taking $\eta(\theta) = p(y \mid \theta)$ in the above.

From a different point of view, McMC sampling can be used for Frequentist parametric inference as follows. Standard asymptotic methods take the curvature of the log-likelihood function at just one point, the peak (or a peak), and use it to estimate the covariance matrix of the parameters and hence make the usual inferences. This is fine if the log-likelihood shape is quadratic, but this may be far from the case in practice, particularly with small samples. A long tail or a multimodal or unbounded likelihood can seriously upset such calculations. Alternatively, the covariance matrix can be estimated from the samples produced by McMC. Moreover, these are generated from the whole of the likelihood function, not just the vicinity of the peak. So, one should obtain a more reliable estimate, and one that will be valid for non-quadratic log-likelihoods. How useful variances and covariances are for non-normal distributions is arguable, but one can also estimate quantiles (as sample proportions) to assess distributional shape and provide alternative bases for inference.

2.4 *Some examples*

In this section, some numerical examples are presented. Complete, definitive analyses are not catalogued — that would require a whole report for each one. Rather, the aim is to illustrate the various methods described above showing the sorts of things that can be done and how.

Example 2.4 Nelson (1970)

Data set 2 of Section 1.2 will be fitted with the exponential mixture model of Example 1.2: the exponential distribution has often been used to describe the failure times of electrical components. Following Lawless (1982, Section 10.1.2), we will focus on cause 9, relabelled as cause 1, with all other failure causes relabelled as cause 2.

The model has three independent parameters, π, with $\pi_1 = \pi$ and $\pi_2 = 1 - \pi$, and λ_1 and λ_2. To fit it, a Fortran program to implement a quasi-Newton method (Appendix 1) was used to locate the maximum point of the log-likelihood function. The latter was returned by sub-routines (available, guarantee-free, on request) set up to generate a more general Weibull mixture, for which the sub-survivor functions are

$$\bar{F}(j, t) = \pi_j \exp\{-(\lambda_j t)^{\phi_j}\} \tag{2.4.1}$$

for $j = 1, \ldots, p$. For the exponential mixture, the ϕ-parameters are fixed at value 1.0. The constraint that the π_js are positive with sum 1 is conveniently ensured by expressing π_j as $\exp(\theta_j)/\sum_{j=1}^{p}\exp(\theta_j)$; in addition, we take $\theta_p = 0$ to ensure identifiability of the θ_j. Such transformations tend to oil the computational wheels by ensuring that parametric combinations cannot stray out of bounds during iterative searches. Incidentally, the observed failure times were scaled down by a factor of 1000, also to smooth the computations; the desired effect is to make the parameter values comparable, the range 0.1 to 1.0 for all of them being an ideal. Again, incidentally, the exponents in Equation 2.4.1 could be written as $-\lambda_j t^{\phi_j}$, rather than $-(\lambda_j t)^{\phi_j}$, but in that case λ_j would lose its status as the scale parameter for t.

The maximum likelihood estimate is $\hat{\theta} = (0.199, 0.220, 0.723)$, which gives $\hat{\pi} = 0.550$, $\hat{\lambda}_1 = 0.220$, and $\hat{\lambda}_2 = 0.723$. The inverse Hessian matrix, of second derivatives of the log-likelihood function, at this point is

$$H^{-1} = 10^{-2}\begin{pmatrix} 11.473 & -0.021 & 0.245 \\ -0.021 & 0.288 & -0.025 \\ 0.245 & -0.025 & 3.560 \end{pmatrix}.$$

This provides an estimate of the covariance matrix of the maximum likelihood estimates, so the standard errors are 0.339 for $\hat{\theta}_1$, 0.054 for $\hat{\theta}_2$, and 0.189 for $\hat{\theta}_3$. That $\hat{\lambda}_2$ has lower precision than $\hat{\lambda}_1$ draws the eye. There are 17 type-1 (formerly type-9) failures, 16 of type-2, and 3 censored failure times. So, the imprecise estimate does not arise from under-representation of type-2 failures in the sample. However, the observed type-1 failure times tend to be larger (which is why $\hat{\lambda}_1 < \hat{\lambda}_2$) and so give a poorer estimate of the corresponding rate. (In a single random sample of size n from an exponential distribution of rate λ, the asymptotic variance of $\hat{\lambda}$ is λ^2/n.)

An approximate 95% confidence interval for θ_1 is $0.199 \pm 1.96 \times 0.339 = (-0.465, 0.863)$, which converts to an interval (0.386, 0.703) for π. Since the latter comfortably covers the value $\pi = 0.5$, equal probabilities of eventual failures of types 1 and 2 cannot be ruled out. Again, the hypothesis of equal rates, $H_0 : \lambda_1 = \lambda_2$, is not obviously supported by the estimates and their standard errors. Formally, for illustration, a Wald test may be applied as follows. The estimated variance of the estimated difference in rates, $\hat{\lambda}_1 - \hat{\lambda}_2$, is

$$10^{-2}\{0.288 - 2 \times (-0.025) + 3.560\} = 0.0390,$$

so the standard normal deviate for testing H_0 is

$$(0.22 - 0.72)/\sqrt{(0.0390)} = 2.53;$$

the corresponding p-value is $p = 0.0114$ so the evidence for different rates is quite strong.

A likelihood ratio test for H_0 can be made by comparing the maximised log-likelihood under the fitted model, here -88.037932, with that fitted under the constraint $\lambda_1 = \lambda_2$, -92.194475. Thus, $2 \times (92.194475 - 88.037932) = 8.31$ can be referred to the chi-square distribution with 1 degree of freedom, giving

$0.001 < p < 0.005$. In theory, 8.31 should equal the square of the aforesaid normal deviate, but this is only an asymptotic, large-sample, equivalence.

The exponential mixture model can be extended to the full Weibull version, Equation 2.4.1. This extended model yields maximum likelihood estimates

$$(\hat{\theta}_1, \hat{\lambda}_1, \hat{\lambda}_2, \hat{\phi}_1, \hat{\phi}_2) = (0.499, 0.261, 0.494, 2.07, 0.624)$$

and maximised log-likelihood –83.279247. A log-likelihood ratio test to compare this fit with the former exponential one yields

$$2 \times (88.037932 - 83.279247) = 9.52$$

to be referred to the chi-square distribution with 2 degrees of freedom. The result, $0.005 < p < 0.01$, implies that the Weibull fit is better, i.e., that at least one of ϕ_1 and ϕ_2 is not equal to 1. We will not pursue this further: the data set is limited, so more searching tests for model-fit are not likely to be productive, and we have bigger fish to fry. However, the example has served to illustrate our basic approach to fitting and testing models. ■

Example 2.5 Hoel (1972)

The data presented as Data Set 4 in Section 1.2 will be examined now. The main issue is the possibility of differences between the two groups. Thus, if we adopt the Weibull mixture model, referred to in the previous example, we will need to allow for possible differences between groups in the π_j, λ_j, and ϕ_j. The full parameter set is then $\theta = (\theta_1, \ldots, \theta_{16})$, in which (π_1, π_2, π_3) is determined by (θ_1, θ_2) for group 1, via the transformation described in Example 2.4, and by (θ_3, θ_4) for group 2; also, $(\lambda_1, \lambda_2, \lambda_3)$ is given by $(\theta_5, \theta_6, \theta_7)$ for group 1 and by $(\theta_8, \theta_9, \theta_{10})$ for group 2, and (ϕ_1, ϕ_2, ϕ_3) is given by $(\theta_{11}, \theta_{12}, \theta_{13})$ for group 1 and by $(\theta_{14}, \theta_{15}, \theta_{16})$ for group 2. The full model, which gives maximised log-likelihood –94.52742, has 16 parameters, but we would hope to reduce it somewhat as described below.

In an univariate Weibull distribution, with survivor function $\exp\{-(\lambda t)^{\phi}\}$, $\log\lambda$ is a location parameter and ϕ is a scale parameter for $\log T$. Therefore, by analogy with normal linear models, let us see first whether the scale parameters can be assumed to be homogeneous over groups. The resulting log-likelihood, maximised under the constraint that (ϕ_1, ϕ_2, ϕ_3) is the same for the two groups, is –104.85453. Similarly, we can fit the model with both the λ and ϕ parameters homogeneous between groups, and likewise for other combinations. Table 2.1 below gives the various maximised log-likelihoods together with the corresponding likelihood ratio chi-squares when tested against the full model. Here, the column $n\pi$ gives the number of distinct π-vectors fitted, so $n\pi = 1$ means homogeneity of π over groups and $n\pi = 2$ means distinct values for the two groups; likewise, columns $n\lambda$ and $n\phi$.

Table 2.1 Tests for Sub-Models vs. Full Model

$n\pi$	$n\lambda$	$n\phi$	log-likelihood	chi-square	df	p-value
2	2	2	–94.52742	—	—	—
2	2	1	–104.85453	20.65	3	$p < 0.001$
2	1	2	–113.39474	37.73	3	$p < 0.001$
1	2	2	–99.37663	9.70	2	$0.005 < p < 0.01$
2	1	1	–127.47499	65.90	6	$p < 0.001$
1	2	1	–109.70377	30.35	5	$p < 0.001$
1	1	2	–118.24399	47.43	5	$p < 0.001$
1	1	1	–132.32424	75.59	8	$p < 0.001$

The first three chi-square tests in Table 2.1 indicate that all three parameter sets, π, λ, and ϕ, appear to differ between groups; the remaining four fits simply confirm this. As far as model reduction is concerned, then, it looks like we are out of luck. However, we have identified differences between groups and should now try to interpret them.

The full maximum likelihood estimates, based on the survival times divided by 1000, are as follows: for group 1, $(\hat{\theta}_1, \hat{\theta}_2) = (-0.573, -0.026)$, yielding $\hat{\pi} = (0.222, 0.384, 0.394)$, $(\hat{\lambda}_1, \hat{\lambda}_2, \hat{\lambda}_3) = (3.206, 1.543, 2.165)$, and $(\hat{\phi}_1, \hat{\phi}_2, \hat{\phi}_3) = (3.591, 8.048, 2.073)$; for group 2, $(\hat{\theta}_3, \hat{\theta}_4) = (-0.270, -0.930)$, giving $\hat{\pi} = (0.354, 0.183, 0.463)$, $(\hat{\lambda}_1, \hat{\lambda}_2, \hat{\lambda}_3) = (2.559, 1.331, 1.262)$, and $(\hat{\phi}_1, \hat{\phi}_2, \hat{\phi}_3) = (2.205, 6.130, 4.393)$. The main differences seem to be a lower probability of death from cause 2 in group 2, and longer average survival times when the eventual causes are 1 or 3. The ϕ-parameters determine finer aspects of the shape of the survival distributions; for instance, as ϕ increases beyond 1.0 in an ordinary Weibull distribution, the density becomes more symmetric.

What should have been done earlier, but was omitted for tidiness, are the routine goodness-of-fit assessments. The conclusions above will not be useful if based on an inappropriate model. Uniform residuals can be defined as described in Section 2.2.1. There are no censored times here so the e_{ij} are defined unambiguously. Figure 2.1 shows plots of the residuals both for individual groups and causes and overall for the two groups. The horizontal and vertical scales here are the expected and observed uniform residuals, respectively. Three of the lines for individual causes have a jump at around 0.4 on the vertical axis, indicative of a gap in the observed residuals, but the numbers of points are a little small for definitive assessment. The two group-plots at the bottom look reasonable in that the uniform residuals follow the ideal 45° line fairly well except for a possible anomaly at the upper end. However, Moran tests (Section 2.2), which are based on the spacings of the uniform residuals, do not draw attention to any of these plots as being out of line. ■

Example 2.6 Lagakos (1978)

The data arose from a lung cancer clinical trial: there are 194 cases with 83 deaths from cause 1 (local spread of disease) and 44 from cause 2 (metastatic

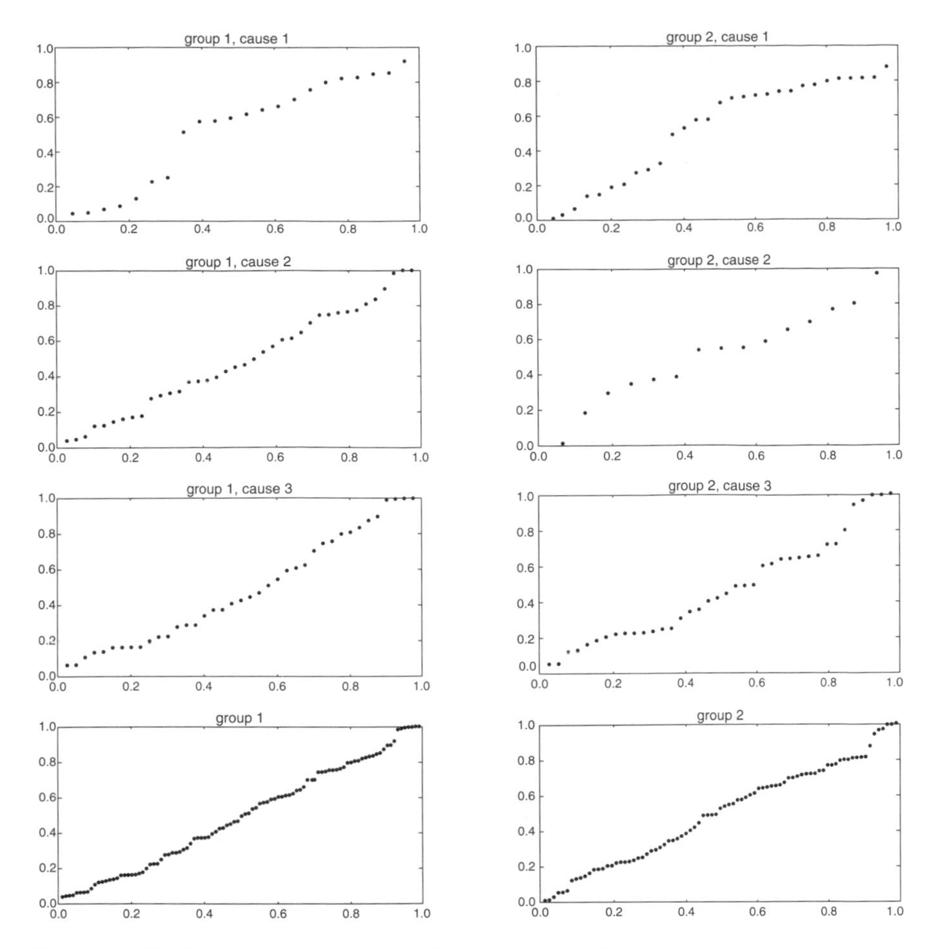

Figure 2.1 Hoel data: uniform residuals by group and cause.

spread), the remaining 67 times being right-censored (cause 0). Three explanatory variables were also recorded: activity performance, a binary indicator of physical state (x_1), a binary treatment indicator (x_2), and age in years (x_3).

Lagakos (1978) listed the data to which he fitted a model with sub-hazard functions of form $h(j, t) = \lambda_j$ $(j = 1, 2)$. Three explanatory variables were accommodated via a log-linear regression model, $\log\lambda_j = x^T\beta_j$ in which $\beta_j = (\beta_{0j}, \beta_{1j}, \beta_{2j}, \beta_{3j})^T$ and x has first component 1 so that β_{0j} is the intercept term. Maximum likelihood estimation was applied, goodness-of-fit was assessed from residual plots, and various hypothesis tests were performed.

Lagakos' model is a special case of that given in Example 1.3, which we will now adopt. Thus, the sub-hazards are $h(j, t) = \phi_j\lambda_j^{\phi_j}t^{\phi_j - 1}$. Fitting the constrained model, in which $\phi_j = 1$ $(j = 1, 2)$, gives the same results as those quoted by Lagakos, with maximised log-likelihood -45.570197. For the unconstrained, full-model, version we obtain –35.211549. The resulting log-likelihood ratio chi-square value, 20.72 (with 2 degrees of freedom, $p < 0.001$), indicates that the constraint is unreasonable. The ϕ-estimates are both greater than 1, giving sub-densities that are zero at $t = 0$ and so are humped rather than being of monotone decreasing, negative exponential shape. Presumably, observed times are too thin on the ground near $t = 0$ for an exponential distribution.

The estimated values of ϕ_1 and ϕ_2 in the full model are quite close, which suggests that the underlying values might be equal. This hypothesis turns out to be tenable: the maximised log-likelihood under this constraint is –35.292728, giving chi-square 0.16 (1 degree of freedom, $p > 0.10$) in comparison with the full model. Thus, we now adopt this model, with ϕ_1 and ϕ_2 both equal to ϕ, say. For convenience we will refer to this as Model 1, the previous, full, unconstrained model being Model 0. The estimates and their standard errors, based on the data in which the times have been scaled by a factor of 0.01, are as follows (the figures fully quoted to enable computation-checking):

$$\begin{aligned}
\hat{\beta}_1 &= (1.2756, 0.46704, 0.31821, -1.2624), \\
\text{se}(\hat{\beta}_1) &= (0.503, 0.169, 0.202, 0.836), \\
\hat{\beta}_2 &= (-0.49924, 0.42610, 0.32170, 0.96176), \\
\text{se}(\hat{\beta}_2) &= (0.751, 0.230, 0.280, 1.18), \\
\hat{\phi} &= 1.3722, \text{se}(\hat{\phi}) = 0.0874.
\end{aligned}$$

As pointed out in Section 1.3, Model 1 has proportional hazards: in this case $\bar{F}(t) = \exp(-st^{\phi})$, where $s = \sum_{j=1}^{p} \lambda_j^{\phi}$, and $\bar{F}(j, t) = \pi_j \bar{F}(t)$, where $\pi_j = \lambda_j^{\phi}/s$.

Further questions of interest, centred on the values of the regression coefficients, β, were considered by Lagakos. Judging by the standard errors given above, one or more of the βs might be zero, in which case the corresponding explanatory variable has no effect on the λ_j value. In fact, leaving aside the intercept terms, β_{01} and β_{02}, with only one exception none of the normal deviate ratios ($\hat{\beta}$-value/standard error) exceeds the magic threshold, 1.96; the exception is $\hat{\beta}_{11}$, for which the ratio is 0.46704/0.169. As a first shot at model reduction, let us re-fit with β_{21}, β_{31}, β_{22}, and β_{32} all set to zero. This yields Model 2 and produces maximised log-likelihood –38.644139; on comparison with Model 1 (proportional hazards), $\chi_4^2 = 6.70$ ($p > 0.10$). So, it seems that the evidence for influence of the second and third x-variables is weak. The parameter estimates for this reduced model are:

$$\begin{aligned}
\hat{\beta}_1 &= (0.80266, 0.37906), \text{se}(\hat{\beta}_1) = (0.107, 0.165), \\
\hat{\beta}_2 &= (0.33379, 0.37738), \text{se}(\hat{\beta}_2) = (0.153, 0.227), \\
\hat{\phi} &= 1.3516, \text{se}(\hat{\phi}) = 0.0854.
\end{aligned}$$

The estimates for these remaining parameters have not undergone drastic revision from the previous values, which is reassuring. Whether we should retain β_{12} is a matter for discussion. Its normal deviate ratio, 0.37738/0.227 = 1.66, is not large. On the other hand, $\hat{\beta}_{11}$ and $\hat{\beta}_{12}$ look remarkably similar. In the parameter covariance matrix the estimates are 273.474×10^{-4} for $\text{var}(\hat{\beta}_{11})$, 515.368×10^{-4} for $\text{var}(\hat{\beta}_{12})$, and 0.761×10^{-4} for $\text{cov}(\hat{\beta}_{11}, \hat{\beta}_{12})$; thus, the estimated variance of the difference, $\hat{\beta}_{11} - \hat{\beta}_{12}$, is

$$10^{-4}(273.474 - 2 \times 0.761 + 515.368) = 0.079.$$

This yields normal deviate $(0.37906 - 0.37738)/\sqrt{(0.079)} = 0.006$. I would not wish to appear for the prosecution of the case against equality here.

The above analysis encourages the fitting of Model 3, incorporating the further constraint $\beta_{12} = \beta_{11}, = \beta$, say. The resulting maximised log-likelihood is –38.644157, almost the same as that from Model 2. The parameter estimates and their standard errors are

$$\hat{\beta}_{01} = 0.80290,\ \text{se}(\hat{\beta}_{01}) = 0.100,\ \hat{\beta}_{02} = 0.33334,\ \text{se}(\hat{\beta}_{02}) = 0.133,$$
$$\hat{\beta} = 0.37848,\ \text{se}(\hat{\beta}) = 0.134,\ \hat{\phi} = 1.3516,\ \text{se}(\hat{\phi}) = 0.0854.$$

These values are almost the same as before but with slightly reduced standard errors. Adopting Model 3, we can say that the effect of x_1 (patient's ability to walk) appears to be real, and the same for the two causes of death, but any effects that x_2 (treatment) and x_3 (age) might have do not show up strongly in the data. These conclusions are in line with those of Lagakos.

Detailed inferences can be extracted via the Bayesian approach. For instance, suppose that the hazard ratio, $h(1, t)/h(2, t)$, the relative risk of failure from the two causes, were at issue. From our fitted model, $h(j, t) = \phi\lambda_j^{\phi}t^{\phi-1}$ with $\log\lambda_j = \beta_{0j} + \beta x_1$, this has logarithm

$$\log\{(\lambda_1/\lambda_2)^{\phi}\} = \phi(\beta_{01} - \beta_{02})$$

independent of β, t, and the x-variables. A modest Markov chain Monte Carlo sampling run, as described in Appendix 2, was performed. First, a Gibbs-like run of length 1000 was used to set up a suitable parameter covariance matrix, and then every tenth sample was retained from a full-dimensional run of total length 10,000 using a Metropolis random walk with a normal jumping distribution. The 1000 retained values produced the following statistics for the log-hazard ratio: mean 0.645, standard deviation 0.192, and quantiles 0.208 (1%), 0.339 (5%), 0.514 (25%), 0.645 (50%), 0.779 (75%), 0.969 (95%), 1.098 (99%). The sampled values and kernel-smoothed posterior density estimates of the individual parameters and of the log-hazard ratio are shown in Figure 2.2. ■

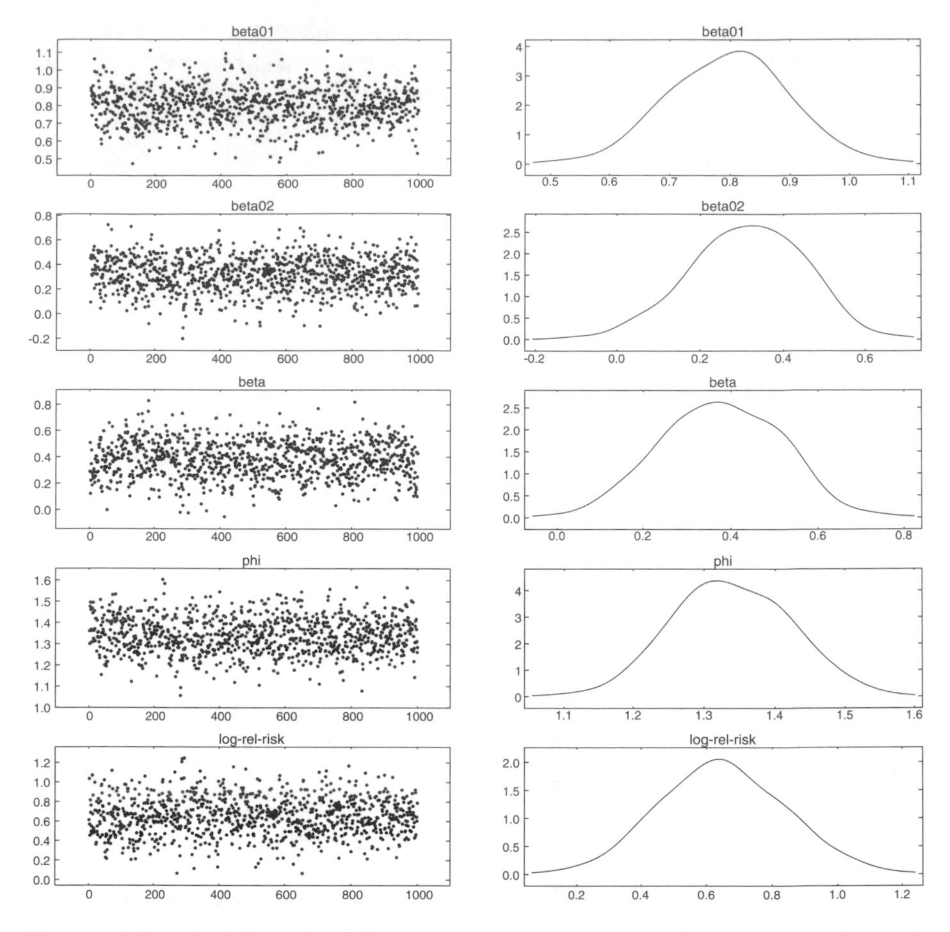

Figure 2.2 Lagakos data: McMC runs — sample paths and histograms for each parameter.

2.5 Masked systems

In many engineering systems, the components are grouped into modules. It is common practice nowadays for repairs to be performed simply by replacing a failed module without bothering to identify the failed component within the module. This is said to be more economical. (But, for whom? I cannot be alone in the experience of trying to replace a very minor component on my car and being told that I have to buy the whole "sealed unit" at vast expense.) In such cases, the failed component will only be identified as one of several. In addition to the work reviewed in more detail below, relevant papers include those by Gross (1970), Miyakawa (1984), Usher and Hodgson (1988), Usher and Guess (1989), Doganaksoy (1991), Guess et al. (1991), and Lin and Guess (1994).

In a parametric analysis involving masked systems, the likelihood can be simply modified as described in Section 2.1.2(a). However, it is clear that if two components, say j_1 and j_2, always appear together in a module, their separate contributions will not be estimable. This is because in the likelihood function Equation 2.1.1, for instance, the sub-density function $f(c_i, t_i)$ will be replaced by $\sum_{j \in c_i} f(j, t_i)$, with summation over the components j in module c_i, and $F(t_i)$ will be expressible as $\sum_{j=1}^{p} \bar{F}(j, t_i)$, with

$$\bar{F}(j, t_i) = \int_{t_i}^{\infty} \sum_{j \in c_i} f(j, t_i);$$

thus, $f(j_1, t)$ and $f(j_2, t)$ only appear together in combination as $f(j_1, t) + f(j_2, t)$ and are therefore separately unidentifiable. If the separate hazards of the different components are to be estimated, then, we need either sufficient distribution of them around different modules, so that the combinations can be disentangled, or some assumptions about the sub-densities.

Albert and Baxter (1995) showed how to apply the EM algorithm to maximise the likelihood for such situations, taking the identifiers of the failing components to be missing data. They mentioned the identifiability problem when two components always appear together in the same module, but said that this was "unlikely in practice."

Reiser et al. (1995) treated a situation in which a subset of components can be identified post-mortem that contains the guilty component, implicitly assuming that "disentanglement" is possible in the data. For illustration, they gave some three-component data in which all six non-empty subsets of {1, 2, 3} appeared in the list of identified failure causes. Such data arise when the breakdown can be diagnosed better in some failed units than in others. The authors were careful to make explicit the assumption mentioned briefly in Section 2.1.2(a), i.e., that the observed time T and set S of possible causes gives no additional information about which individual cause C in S was the source of the trouble: in probability terms, their assumption is that, for some fixed $j' \in s$,

$$\text{pr}(S = s \mid C = j, T = t) = \text{pr}(S = s \mid C = j', T = t) \text{ for all } j \in s. \quad (2.5.1)$$

This would not be true, for instance, in a two-component system where complete destruction of the system, and, with it, any causal evidence, is more likely to be due to failure of the second component than of the first.

The contribution of a failure time t with diagnosed cause-subset s to the likelihood function is, under Equation 2.5.1,

$$\begin{aligned} \text{pr}(S = s, T = t) &= \sum_{j \in s} \text{pr}(C = j, T = t)\ \text{pr}(S = s \mid C = j, T = t) \\ &= \text{pr}(S = s \mid C = j', T = t) \sum_{j \in s} f(j, t), \end{aligned}$$

writing $f(j, t)$ for $\text{pr}(C = j, T = t)$; the application of Equation 2.5.1 removes the terms $\text{pr}(S = s \mid C = j, T = t)$ and replaces them with the single factor $\text{pr}(S = s \mid C = j', T = t)$ outside the summation. Hence, the likelihood function Equation 2.1.1 becomes

$$L = \prod_{o} \{\text{pr}(S = s_i|C = j', T = t_i) \sum_{j \in s_i} f(j, t_i)\} \times \prod_{c} \bar{F}(t_i).$$

Reiser et al. pointed out that, provided that the factor $\Pi_o \text{pr}(S = s_i \mid C = j', T = t_i)$ does not involve the parameters of interest, it may be treated effectively as a constant and omitted from the calculations. They went on to present a Bayesian analysis in which, for simplicity, the component lifetimes are independent exponential variates, and using the usual invariant improper prior for the exponential rate parameters.

The approach outlined in the previous paragraph was extended by Guttman et al. (1995) to the case of "dependent masking" in which Equation 2.5.1 is replaced by

$$\text{pr}(S = s \mid C = j, T = t) = a_j \, \text{pr}(S = s \mid C = j', T = t) \text{ for all } j \in s, \quad (2.5.2)$$

where $j' \in s$ is fixed and a_j does not depend on t. The assumption is that the relative likelihoods of masking, given different component failures, do not change over time. It reduces to Equation 2.5.1 when $a_j = 1$ for all j; if $a_j = 0$, failure of component j cannot yield the observation $S = s$. Under Equation 2.5.2, the likelihood contributions have the form

$$\begin{aligned} \text{pr}(S = s, T = t) &= \sum_{j \in s} \text{pr}(C = j, T = t) \; \text{pr}(S = s|C = j, T = t) \\ &= \text{pr}(S = s|C = j', T = t) \sum_{j \in s} a_j f(j, t). \end{aligned}$$

Guttman et al. considered a two-component system, so there is only one quantity a_j to deal with. They applied a Bayesian analysis with component lifetimes and prior as in Reiser et al. (1995).

chapter 3

Latent failure times: probability distributions

The specification of Competing Risks via a set of latent lifetimes is covered in this chapter. This is the traditional route, often with the additional assumption of independence between these concealed lifetimes ("born to blush unseen, and waste their sweetness on the desert air," etc.). It should be mentioned at the outset that this approach carries a health warning: there is a certain problem of model identifiability that will be examined in detail in Chapter 7.

3.1 Basic probability functions

3.1.1 Multivariate survival distributions

Let $\boldsymbol{T} = (T_1, \ldots, T_p)$ be a vector of failure times. In the Competing Risks context, the T_j will have a particular interpretation, but in this preliminary subsection, we just treat them as a set of non-negative random variables. The **joint distribution function** is defined as $G(\boldsymbol{t}) = \text{pr}(\boldsymbol{T} \leq \boldsymbol{t})$, where $\boldsymbol{t} = (t_1, \ldots, t_p)$ and $\boldsymbol{T} \leq \boldsymbol{t}$ means $T_j \leq t_j$ for each j; similarly, the **joint survivor function** is $\bar{G}(\boldsymbol{t}) = \text{pr}(\boldsymbol{T} > \boldsymbol{t})$. If the T_j are jointly continuous, their **joint density** is $\partial^p G(\boldsymbol{t})/\partial t_1 \ldots \partial t_p$, or, equivalently, $(-1)^p \partial^p \bar{G}(\boldsymbol{t})/\partial t_1 \ldots \partial t_p$.

Independence of the T_j is defined equivalently by

$$G(\boldsymbol{t}) = \Pi_{j=1}^{p} G_j(t_j) \quad \text{or} \quad \bar{G}(\boldsymbol{t}) = \Pi_{j=1}^{p} \bar{G}_j(t_j),$$

where $G_j(t_j) = \text{pr}(T_j \leq t_j)$ is the **marginal distribution function** of T_j and $\bar{G}_j(t_j) = \text{pr}(T_j > t_j)$ is the **marginal survivor function** of T_j. When the T_j are not independent the degree of dependence can be measured in various ways. Two such closely related measures are the ratios

$$R(\boldsymbol{t}) = G(\boldsymbol{t})/\Pi_{j=1}^{p} G_j(t_j), \text{ and } \bar{R}(\boldsymbol{t}) = \bar{G}(\boldsymbol{t})/\Pi_{j=1}^{p} \bar{G}_j(t_j);$$

if $R(\boldsymbol{t}) > 1$, or $\bar{R}(\boldsymbol{t}) > 1$, for all $\boldsymbol{t}$, there is positive dependence and vice versa (e.g., Shaked, 1982). For example, for a bivariate distribution $\bar{R}(\boldsymbol{t}) < 1$ if and only if $\text{pr}(T_1 > t_1 \mid T_2 > t_2) < \text{pr}(T_1 > t_1)$, i.e., T_1 becomes less likely to exceed t_1 after it becomes known that $T_2 > t_2$.

3.1.2 *Latent lifetimes*

In the traditional approach to Competing Risks it is assumed that there is a potential failure time associated with each of the p risks to which the system is exposed. Thus, T_j represents the time to system failure from cause j ($j = 1, \ldots, p$): the smallest T_j determines the time T to overall system failure, and its index C is the cause of failure, i.e., $T = \min\{T_1, \ldots, T_p\} = T_C$. Once the system has failed, the remaining lifetimes are lost to observation — in a sense they cease to exist, if they ever did. For example, in an electric circuit with p components in series the circuit will be broken as soon as any one component fails: T is then equal to the failure time of the component that fails first, and C identifies the failing component.

The vector $\boldsymbol{T} = (T_1, \ldots, T_p)$ will have a joint survivor function $\bar{G}(\boldsymbol{t}) = \text{pr}(\boldsymbol{T} > \boldsymbol{t})$ and marginal survivor functions $\bar{G}_j(t) = \text{pr}(T_j > t)$. In the older, actuarial and demographic, terminology the sub-survivor functions $\bar{F}(j, t)$ were called the **crude survival functions** and the marginal $\bar{G}_j(t)$ functions were called the **net survival functions**. It will be assumed for the present that the T_j are continuous and that ties cannot occur, i.e., $\text{pr}(T_j = T_k) = 0$ for all $j \neq k$, otherwise C is not so simply defined (Section 7.2). If the joint survivor function $\bar{G}(\boldsymbol{t})$ of $\boldsymbol{T}$ has known form, then $\bar{F}(t)$ can be evaluated as $\bar{G}(t\mathbf{1}_p)$, where $\mathbf{1}_p = (1, \ldots, 1)$ is of length p.

Theorem 3.1 (Tsiatis, 1975)

The sub-densities can be calculated directly from the joint survivor function of the latent failure times as

$$f(j, t) = [-\partial\bar{G}(\boldsymbol{t})/\partial t_j]_{t\mathbf{1}_p}, \tag{3.1.1}$$

the notation $[\ldots]_{t\mathbf{1}_p}$ indicating that the enclosed function is to be evaluated at $\boldsymbol{t} = t\mathbf{1}_p = (t, \ldots, t)$.

Proof

By definition,

$$\begin{aligned} f(j, t) &= \lim_{\delta \to 0}\delta^{-1}\{\bar{F}(j, t) - \bar{F}(j, t+\delta)\} \\ &= \lim_{\delta \to 0}\delta^{-1}\text{pr}\{t < T_j \leq t+\delta, \cap_{k \neq j}(T_k > T_j)\}. \end{aligned}$$

But,

$$\text{pr}\{t < T_j \le t + \delta, \cap_{k \ne j}(T_k > t + \delta)\} \le \text{pr}\{t < T_j \le t + \delta, \cap_{k \ne j}(T_k > T_j)\}$$
$$\le \text{pr}\{t < T_j \le t + \delta, \cap_{k \ne j}(T_k > t)\}.$$

Divide by δ and then let $\delta \to 0$ to obtain the result. ■

It follows immediately from the theorem that the sub-hazards can also be calculated directly from $\bar{G}(\boldsymbol{t})$ as

$$h(j, t) = f(j, t)/\bar{F}(t) = [-\partial \log \bar{G}(t)/\partial t_j]_{t1_p}. \tag{3.1.2}$$

Example 3.1 Bivariate exponential (Gumbel, 1960)

Gumbel suggested three different forms for bivariate exponential distributions. We will use the first of these for illustration at various points throughout this book on account of its simplicity. It has been pointed out elsewhere that it does not have any obvious justification as describing a realistic physical mechanism. This total lack of credibility is part of its appeal. It has joint survivor function

$$\bar{G}(\boldsymbol{t}) = \exp(-\lambda_1 t_1 - \lambda_2 t_2 - \nu t_1 t_2),$$

in which $\lambda_1 > 0$ and $\lambda_2 > 0$; the dependence parameter ν satisfies $0 \le \nu \le \lambda_1\lambda_2$, $\nu = 0$ yielding independence of T_1 and T_2. The dependence ratio $\bar{R}(t)$ (Section 3.1.1) is equal to $\exp(-\nu t_1 t_2)$ here, and since this is always smaller than 1, the distribution exhibits only negative dependence.

The marginal survivor function of T is $\bar{F}(t) = \exp(-\lambda_+ t - \nu t^2)$, where $\lambda_+ = \lambda_1 + \lambda_2$. For $j = 1, 2$, we have the sub-densities

$$f(j, t) = [(\lambda_j + \nu t_{3-j})\bar{G}(\boldsymbol{t})]_{(t,t)} = (\lambda_j + \nu t)\bar{F}(t)$$

and the sub-survivor functions

$$\begin{aligned}
\bar{F}(j, t) &= \int_t^\infty (\lambda_j + \nu s)\exp(-\lambda_+ s - \nu s^2)ds \\
&= \frac{1}{2}[-\exp(-\lambda_+ s - \nu s^2)]_t^\infty + \frac{1}{2}(\lambda_j - \lambda_{3-j})\int_t^\infty \exp(-\lambda_+ s - \nu s^2)ds \\
&= \frac{1}{2}\exp(-\lambda_+ t - \nu t^2) \\
&\quad + \frac{1}{2}(\lambda_j - \lambda_{3-j})(\pi/\nu)^{1/2}\exp(\lambda_+^2/4\nu)\Phi\{-(2\nu)^{1/2}(t + \lambda_+/2\nu)\},
\end{aligned}$$

where Φ is the standard normal distribution function. We seem to have started with an exponential distribution and ended up with the normal — Competing Risks is full of surprises! If $\lambda_1 = \lambda_2$ we have the simplified form

$$\bar{F}(j, t) = \frac{1}{2}\exp(-\lambda_+ t - \nu t^2).$$

The sub-hazard functions are $h(j, t) = \lambda_j + \nu t$, and the overall system hazard function is $h(t) = \lambda_+ + 2\nu t$. If $\nu = 0$, the latent failure times are independent and we have a constant hazard rate $h(t) = \lambda_+$, as might be expected from exponential distributions. The effect of dependence is to make the overall hazard increase linearly in time, as might not be expected. Proportional hazards obtains if and only if $(\lambda_1 + \nu t)/(\lambda_2 + \nu t)$ is independent of t, i.e., if either $\lambda_1 = \lambda_2$ or $\nu = 0$. ■

As a complement to Theorem 3.1 the sub-densities can be calculated as follows from the joint density $g(\boldsymbol{t})$ of $\boldsymbol{T}$, which is useful for some purposes. Substituting the expression

$$\bar{G}(\boldsymbol{t}) = \int_{t_1}^{\infty} ds_1 \ldots \int_{t_p}^{\infty} ds_p \{g(\boldsymbol{s})\}$$

into the theorem, we obtain

$$f(j, t) = \int_t^{\infty} ds_1 \ldots \int_t^{\infty} ds_{j-1} \int_t^{\infty} ds_{j+1} \ldots \int_t^{\infty} ds_p \{g(\boldsymbol{s})\}; \tag{3.1.3}$$

thus, $g(\boldsymbol{t})$ is integrated over (t, ∞) with respect to each coordinate except the jth. For $p = 2$, it yields $f(j, t) = \int_t^{\infty} g(\boldsymbol{r}_j) ds$, where $\boldsymbol{r}_1 = (t, s)$ and $\boldsymbol{r}_2 = (s, t)$.

Figure 3.1 shows which probabilities are determined by the sub-survivor functions and marginal survivor functions in the bivariate case. Thus, $\bar{F}(2, t)$ and $\bar{G}_2(t)$ determine probabilities in horizontal strips either side of the line $t_1 = t_2$; $\bar{G}_2(t)$ here denotes the marginal survivor function of T_2, by definition $\text{pr}(T_2 > t)$. Similarly, $\bar{F}(1, t)$ and $\bar{G}_1(t)$ determine the content of vertical strips. The integral formula for $f(j, t)$ given above for the bivariate case is illustrated by the probability mass b in Figure 3.1 when $s_2 \to s_1$: $f(2, t) = \int_t^{\infty} g(\boldsymbol{r}) ds$ with $\boldsymbol{r} = (s, t)$.

3.2 Some examples

One or two examples from the literature will be described, and then one or two well-known multivariate failure time distributions will be examined from the point of view of Competing Risks.

Example 3.2 Moeschberger and David (1971)

Moeschberger and David (1971) wrote down the forms of the likelihood functions for generally dependent risks for fully observed failure times and

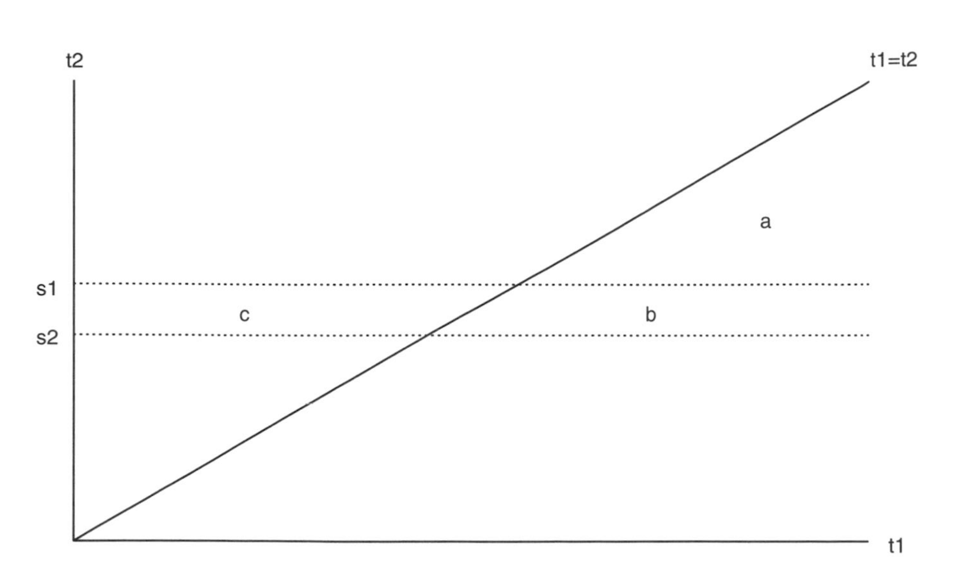

Figure 3.1 Probability content of various regions for bivariate case: a = $\bar{F}(2, s_2)$; b = $\bar{F}(2, s_1) - \bar{F}(2, s_2)$; c = $\bar{G}_2(s_1) - \bar{G}_2(s_2) -$ b.

for grouped and censored failure times. They went on to consider in detail independent Weibull and exponential latent failure times, giving formulae for the likelihood function and its derivatives. Explanatory variables were not considered. They applied the Weibull model to Boag's (1949) data. ■

Example 3.3 Moeschberger (1974)

Moeschberger (1974) derived the forms of the likelihood functions for bivariate normal and Weibull distributions. The former does not yield explicit expressions for the sub-densities, they involve the normal distribution function integral; in any case, the normal distribution does not provide a wholly convincing model since lifetimes cannot be negative and their distributions are usually positively skewed. The latter was taken to be of the Marshall-Olkin type, which will be examined in Section 7.2. Moeschberger discussed various properties of the Weibull model and fitted it, by maximum likelihood, to some of Nelson's (1970) data (Group 5, automatic testing, before the modification on 04.10): three rather extreme (c, t) pairs, (8, 45), (0, 378), and (0, 470), were omitted from the analysis. ■

Ebrahimi (1996) extended parametric modelling with independent latent failure times to the case of uncertainly observed failure causes.

Example 3.4 Bivariate exponential (Freund, 1961)

In a two-component system the lifetimes, T_1 and T_2, are initially independent exponential variates. However, failure of one component alters the subsequent development of the other, the effect being to change the exponential rate parameter of the surviving component from λ_j to μ_j ($j = 1, 2$). The case $\mu_j > \lambda_j$ would be appropriate when the stress on the surviving component is increased after the failure, as when the load was previously shared. Conversely, $\mu_j < \lambda_j$ would reflect some relief of stress, as when the components previously competed for resources.

The joint survivor function can be derived as

$$\bar{G}(\boldsymbol{t}) = (\lambda_+ - \mu_2)^{-1}\{\lambda_1 e^{-(\lambda_+ - \mu_2)t_1 - \mu_2 t_2} + (\lambda_2 - \mu_2)e^{-\lambda_+ t_2}\}$$

for $t_1 \leq t_2$, assuming that $\lambda_+ = \lambda_1 + \lambda_2 \neq \mu_2$; the form for $t_1 \geq t_2$ is obtained from the above by interchanging λ_1 and λ_2, μ_1 and μ_2, t_1 and t_2. Setting $t_1 = 0$ in the form given for $\bar{G}(\boldsymbol{t})$, we obtain the marginal

$$\bar{G}_2(t_2) = (\lambda_+ - \mu_1)^{-1}\{\lambda_1 e^{-\mu_2 t_2} + (\lambda_2 - \mu_2)e^{-\lambda_+ t_2}\}.$$

This is a mixture of two exponentials, rather than a single one, so the joint distribution is not strictly speaking a bivariate exponential. Independence of T_1 and T_2 obtains if and only if $\mu_1 = \lambda_1$ and $\mu_2 = \lambda_2$.

Using Equations 3.1.1 and 3.1.2, the sub-densities, survivor and hazard functions are derived as

$$f(j, t) = \lambda_j e^{-\lambda_+ t},\ \bar{F}(j, t) = (\lambda_j/\lambda_+)e^{-\lambda_+ t},$$
$$\bar{F}(t) = e^{-\lambda_+ t},\ h(j, t) = \lambda_j.$$

Notice that these functions involve (λ_1, λ_2) but not (μ_1, μ_2); this is because they describe only the time to first failure. In consequence, the likelihood function of Section 2.1.1 does not involve (μ_1, μ_2) and so no information is forthcoming on these parameters from standard Competing Risks data. However, full information can be obtained if the system can be observed in operation after a first failure, which is just the situation for which the model is constructed.

The fact that $h(j, t) = \lambda_j$, independent of t, for this distribution should come as no surprise because of the way in which the distribution is set up. During the time before the first failure, which is what the $h(j, t)$ describe, the component failure times are independent exponential variates. During this phase the sub-hazard functions are equal to the marginal ones, and these are constant.

A special case of this model was proposed by Gross et al. (1971). They considered the symmetric case, $\lambda_1 = \lambda_2$ and $\mu_1 = \mu_2$, as appropriate when an individual can survive the loss of one of a pair of organs such as the lungs or kidneys. ■

Example 3.5 Frailty models

To develop this class of models we begin with the marginal survivor function of T_j: $\bar{G}_j(t) = \exp\{-H_j(t)\}$, where $H_j(t)$ is the integrated hazard function of T_j

(Section 1.1.1). Suppose that the H_j ($j = 1, \ldots, p$) for an individual share a common random factor z, and that z varies over the population of individuals with distribution function K on $(0, \infty)$. In some contexts, z is known as the **frailty** of the individual (Vaupel et al., 1979). Replacing $H_j(t)$ by $zH_j(t)$, and assuming that the T_j for an individual are conditionally independent given z, the joint survivor function conditional on z is

$$\bar{G}(\boldsymbol{t}|z) = \exp\{-z\sum_{j=1}^{p} H_j(t_j)\}.$$

The unconditional joint survivor function is now calculated by integrating over the z-distribution as

$$\bar{G}(\boldsymbol{t}) = \int_0^{\infty} e^{-zs} dK(z) = L_K(s), \tag{3.2.1}$$

say, where $s = H_1(t_1) + \ldots + H_p(t_p)$; L_K is the Laplace transform, and moment generating function, of the distribution K. The construction gives a multivariate distribution in which the component lifetimes are dependent, in general. However, the form of dependence is restricted by the symmetry of $\bar{G}$ in the H_j.

In general terms we can now derive $\bar{F}(t)$ from $\bar{G}(t\mathbf{1}_p)$ as $L_K(s_t)$, where $s_t = H_1(t) + \ldots + H_p(t)$, and

$$f(j, t) = [-\partial L_K(s_t)/\partial t_j]_{t\mathbf{1}_p} = -h_j(t)L_K'(s_t),$$

with $L_K'(s) = dL_K(s)/ds$. The sub-hazard functions are then given by

$$h(j, t) = -h_j(t)L_K'(s_t)/L_K(s_t) = -h_j(t)d\log L_K(s_t)/ds_t.$$

We cannot get much further without considering particular cases. For $H_j(t)$, we now take the form $\xi_j t^{\phi_j}$, which is the integrated Weibull hazard, and then $s_t = \Sigma\xi_j t^{\phi_j}$. For K, the following two choices prove to be fruitful.

(a) Multivariate Burr (Takahasi, 1965; Clayton, 1978; Hougaard, 1984; Crowder, 1985)

Taking K to be a gamma distribution with shape parameter ν yields $\bar{G}(\boldsymbol{t}) = (1 + s)^{-\nu}$. The sub-densities are then

$$f(j, t) = \nu\phi_j\xi_j t^{\phi_j - 1}(1 + s_t)^{-\nu - 1},$$

and $\bar{F}(t) = (1 + s_t)^{-\nu}$. Calculation of $\bar{F}(j, t)$, as the integral of $f(j, t)$, is tractable only if $\phi_j = \phi$ for all j; in this case $s_t = \xi_+ t^{\phi}$ and $\bar{F}(j, t) = (\xi_j/\xi_+)(1 + s_t)^{-\nu}$, where $\xi_+ = \Sigma\xi_j$. The sub-hazard functions are given by

$$h(j, t) = \nu\phi_j\xi_j t^{\phi_j - 1}(1 + s_t)^{-1};$$

they are proportional if $\phi_j = \phi$ for all j. The sub-hazards here have a richer structure than in the previous examples. For $\phi_j < 1$, $h(j, t)$ is infinite at $t = 0$ and decreasing thereafter, but for $\phi_j > 1$ its behaviour is less transparent. What is easy to see for $\phi_j > 1$ is that $h(j, t) \sim \nu\phi_j\xi_j t^{\phi_j - 1}$ for small t, and

$$h(j, t) \sim (\nu\phi_j\xi_j/\xi_m)t^{\phi_j - \phi_m - 1}$$

for large t, where m is the index of the largest ϕ_k; thus, $h(j, t)$ is initially increasing from zero at $t = 0$, and eventually behaves as $t^{\phi_j - \phi_m - 1}$ where $\phi_j - \phi_m - 1 \leq -1$.

Figure 3.2(a) shows some plots. In particular, the sub-hazard functions rise at first but eventually fall back to earth. This rather limits the application of this model: in most situations hazard functions tend to increase with age.

(b) Multivariate Weibull (Gumbel, 1960; Watson and Smith, 1985; Hougaard, 1986a, b)

Taking K to be a positive stable distribution with characteristic exponent $\nu \in (0, 1)$ yields $\bar{G}(\boldsymbol{t}) = \exp(-s^\nu)$. The sub-densities are then

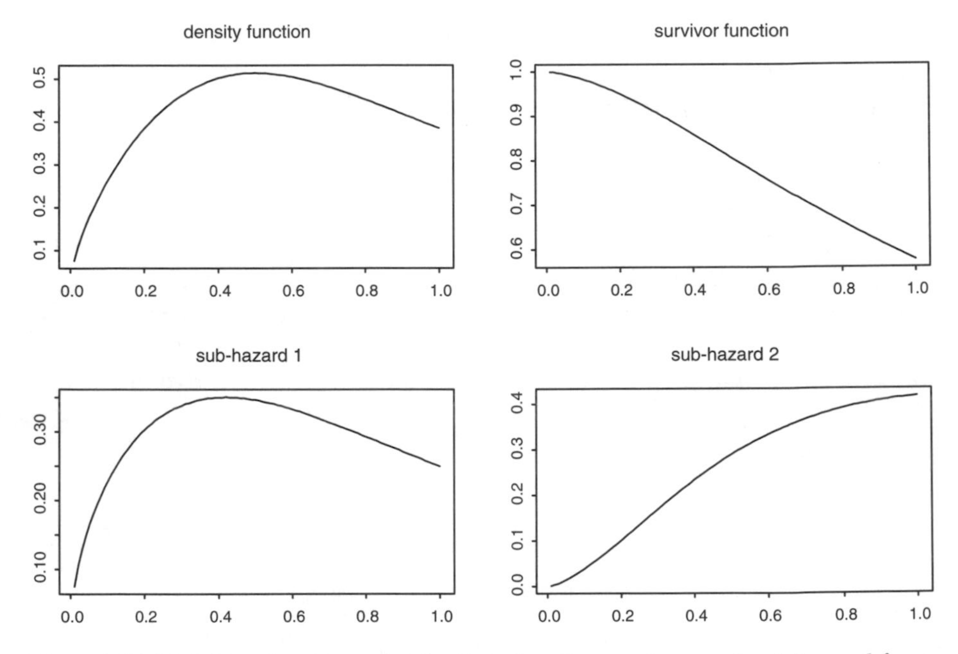

Figure 3.2(a) Multivariate Burr distribution: density, survivor, and sub-hazard functions for $(\nu, \xi_1, \xi_2, \phi_1, \phi_2) = (0.5, 1, 1, 1.5, 2.5)$.

$$f(j, t) = \nu\phi_j\xi_j t^{\phi_j - 1} s_t^{\nu - 1}\exp(-s_t^{\nu}),$$

and $\bar{F}(t) = \exp(-s_t^{\nu})$. Calculation of $\bar{F}(j, t)$, as the integral of $f(j, t)$, is tractable only if $\phi_j = \phi$ for all j; in this case $s_t = \xi_+ t^{\phi}$ and $\bar{F}(j, t) = (\xi_j/\xi_+)\exp(-s_t^{\nu})$. The sub-hazard functions are given by

$$h(j, t) = \nu\phi_j\xi_j t^{\phi_j - 1} s_t^{\nu - 1};$$

they are proportional if $\phi_j = \phi$ for all j. For $\phi_j < 1$, $h(j, t)$ is infinite at $t = 0$ and decreasing thereafter, remembering that $0 < \nu < 1$. For $\phi_j > 1$,

$$h(j, t) = (\nu\phi_j\xi_j/\xi_l^{1-\nu})t^{\phi_j - 1 - (1-\nu)\phi_l}$$

for small t, where ϕ_l is the least among the ϕ_k, and

$$h(j, t) \sim (\nu\phi_j\xi_j/\xi_m^{1-\nu})t^{\phi_j - 1 - (1-\nu)\phi_m}$$

for large t, where ϕ_m is the maximum of the ϕ_j. Thus, for $\phi_j > 1$, $h(j, t)$ behaves like $t^{\phi_j - 1 - (1-\nu)\phi_l}$ near $t = 0$ and like $t^{\phi_j - 1 - (1-\nu)\phi_m}$ as $t \to \infty$, where both $\phi_j - 1 - (1 - \nu)\phi_l$ and $\phi_j - 1 - (1 - \nu)\phi_m$ could be positive or negative. Thus, for example, the fabled bathtub shape (down-along-up) is not attainable because it would require $\phi_j - 1 - (1 - \nu)\phi_l < 0$ and $\phi_j - 1 - (1 - \nu)\phi_m > 0$.

Figure 3.2(b) shows some plots. The sub-hazards here can rise or fall with age, so the model is more flexible in this respect than the multivariate Burr. ■

Farragi and Korn (1996) considered a particular family of frailty models previously suggested by Oakes (1989). For Competing Risks in the clinical context they proposed a version in which the medical treatment affects only one of the latent failure times.

Example 3.6 A stochastic process model

Yashin et al., (1986) proposed the following set-up in connection with human mortality. There is an unobserved physiological ageing process $\{z(t):t \geq 0\}$, determined by a stochastic differential equation:

$$dz(t) = \{a_0(t) + a_1(t)z(t)\}dt + b(t)W(t),$$

where $W(t)$ is a standard Wiener process or Brownian Motion of specified dimension. In the terminology, $z(t)$ is an Ito process with drift coefficient

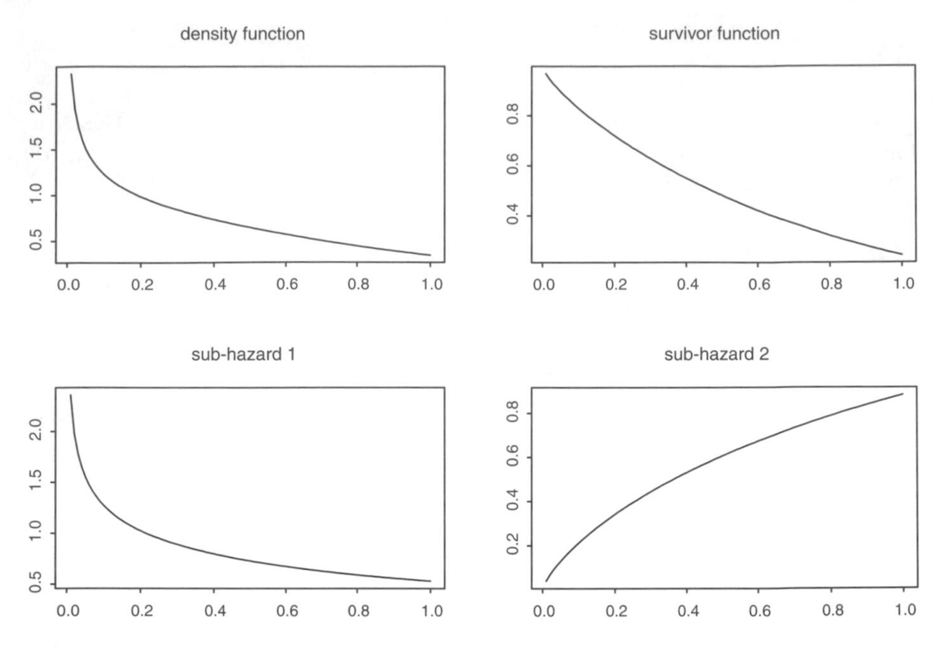

Figure 3.2(b) Multivariate Weibull distribution: density, survivor and sub-hazard functions for $(\nu, \xi_1, \xi_2, \phi_1, \phi_2) = (0.5, 1, 1, 1.5, 2.5)$.

$a_0(t) + a_1(t)z(t)$ and diffusion coefficient $b(t)$, and Yashin et al. give arguments in support of these model specifications for the context. The essence of the Competing Risks model here is that the latent failure times are conditionally independent given the current history, $H_t = \{z(s):0 \leq s \leq t\}$, of the process: specifically,

$$\text{pr}(T > t|H_t) = \Pi_{j=1}^{p} \text{pr}(T_j > t_j|H_t).$$

Further, it is assumed that $\text{pr}(T_j > t_j \mid H_t) = h_j(t_j \mid z_t)$, a marginal hazard depending on H_t only through the current value z_t. To obtain the marginal hazard functions, $h_j(t_j) = d\log G_j(t_j)/dt_j$, z_t must be integrated out in the same way that the random effects, or unobserved frailties, are integrated out in Example 3.5, Equation 3.2.1. The various other functions now follow, and Yashin et al. give some examples, further development, and an application to some mortality data. ■

3.3 *Marginal vs. sub-distributions*

Historically, the main aim of Competing Risks analysis was seen as the estimation of the marginal distributions, i.e., taking data in which the risks act together and trying to infer how some of them would act in isolation. In the older terminology (Section 3.1.2), this is an attempt to make inferences about the **net risks** from observations on the **crude risks**. In more everyday terms, it's like investigative reporting, uncovering private lives or, rather, lifetimes. In algebraic terms it amounts to deriving the $\bar{G}_j(t)$ from the $\bar{F}(j, t)$. It will be seen in Chapter 7 that this cannot be done unless $\bar{G}(t)$ is restricted in some way, perhaps by specifying a fully parametric model for it. Without making such assumptions, the best that can be done is to derive some bounds for $\bar{G}_j(t)$ in terms of the $\bar{F}(j, t)$. The question of whether it is sensible to want to estimate the $\bar{G}_j(t)$ in the first place is another matter. It is only natural to want to focus on the cause of failure of major interest and make inferences and predictions free of the nuisance aspects. For example, the doctor would probably want to give advice about the illness in question, without having to bring in the possibility of being run over by a bus. However, the counter-argument is that the $\bar{G}_j(t)$ do not describe events that physically occur — they only describe failures from isolated causes in situations in which all the other risks have been removed somehow. It is the $\bar{F}(j, t)$, not the $\bar{G}_j(t)$, that are relevant to the real situation, i.e., to failures from cause j that can actually occur. Prentice et al. (1978) emphasized these points in their vigorous criticism of the traditional approach to Competing Risks.

In certain special circumstances, it might be possible to observe the effect of failure cause j alone, i.e., with other causes absent. In that case, the **marginal hazard rate** $h_j(t) = -d\log\bar{G}_j(t)/dt$ is relevant; $h_j(t)$ is also known as the marginal intensity function. However, it might happen that the removal of the other causes of failure materially changes the circumstances. In that case, it is not valid to assume that when T_j is observed in isolation its distribution is the same as the former marginal T_j-distribution derived from the joint distribution. This has long been recognized (e.g., Makeham, 1874 and Cornfield, 1957).

Elandt-Johnson (1976) made a thoughtful distinction in this connection. She differentiated between (i) simply ignoring some causes of failure, in which case one sets the corresponding t_j to zero in $\bar{F}(\mathbf{t})$ and works with the marginal survivor function of the rest, and (ii) eliminating or postponing some causes of failure by improved care or maintenance, in which case one should work with the conditional survivor function of the rest given that the corresponding T_j are effectively infinite. Clearly, (i) and (ii) will in general give different answers and she gave examples to illustrate this.

An obvious inequality connecting the observable and unobservable functions is $\bar{F}(j, t) \leq \bar{G}_j(t)$, i.e., $\mathrm{pr}(C = j, T_j > t) \leq \mathrm{pr}(T_j > t)$. The following theorem gives more refined ones.

Theorem 3.2 (Peterson, 1976)

(i) Let $v = \max\{t_1, \ldots, t_p\}$. Then

$$\sum_{j=1}^{p} \bar{F}(j, v) \leq \bar{G}(\mathbf{t}) \leq \sum_{j=1}^{p} \bar{F}(j, t_j).$$

(ii) For $t > 0$ and each k,

$$\sum_{j=1}^{p} \bar{F}(j, t) \leq \bar{G}_k(t) \leq \bar{F}(k, t) + (1 - p_k).$$

Proof

(i) Note that $\bar{G}(\boldsymbol{t}) = \sum_{j=1}^{p} \text{pr}(C = j, \boldsymbol{T} > \boldsymbol{t})$ and that the right hand side is bounded by:

$$\text{rhs} \geq \sum_j \text{pr}(C = j, T > v\mathbf{1}_p) = \sum_j \text{pr}(C = j, T > v) = \sum \bar{F}(j, v);$$

$$\text{rhs} \leq \sum_j \text{pr}(C = j, T_j > t_j) = \sum_j \bar{F}(j, t_j).$$

(ii) Set $t_k = t$ and all other $t_j = 0$ in (i). ■

Peterson (1976) remarked that, although the bounds cannot be improved, because they are attainable by particular distributions, they are generally rather wide, not restricting $\bar{G}(\boldsymbol{t})$ and $\bar{G}_j(t_j)$ with any great precision. Thus, it's possible to have a pretty good picture of the $\bar{F}(j, t)$ without this picture's telling you much about $\bar{G}(\boldsymbol{t})$ or the $\bar{G}_j(t_j)$. We shall return to this identifiability aspect in Chapter 7. More recently, Bedford and Meilijson (1998) have given more refined bounds for the marginals, and other developments have been made by Nair (1993) and Karia and Deshpande (1997).

Rachev and Yakovlev (1985) studied the inverse problem to Peterson's, deriving bounds for the sub-survivor functions $\bar{F}(j, t)$ in terms of the marginal $\bar{G}_j(t)$. One can envisage a situation where the components can be tested individually in order to predict their behaviour when incorporated into a system. From the marginals they derived bounds for the probabilities $p_j = \bar{F}(j, 0)$ in a two-component system, and for the covariance $\text{cov}(T_1, T_2)$. They applied their bounds in a certain model for the effects and interaction of two injuries, and derived confidence intervals for the p_j. They assumed for this that empirical estimates of the functions $\bar{G}_1(t)$, $\bar{G}_2(t)$, and $\bar{F}(t)$ are available.

3.4 Independent risks

The classical assumption is that the risks act independently, i.e., that the latent failure times T_j are independent. This would appear to be especially dubious in medical applications, which is, ironically, the context in which Competing Risks developed. The complex biochemical interactions between different disease processes and physiological conditions would seem to rule out any hope of independence at the outset, making such an assumption not even worthy of tentative consideration. On the other hand, in reliability theory, a lot of work involving complex systems is based on an assumption of stochastic independence, this being justified by the physically independent functioning of components. However, even here the assumption of independent risks is often not likely to be anywhere near true, particularly when there is load-sharing between components, or randomly fluctuating loads, or other shared common factors such as working environment, previous exposure to wear and tear, and quality of materials, manufacture, and maintenance. To assume independence, one must be sure that a failure of one type has absolutely no bearing at all on the likelihood of failure of any other type, not even through some indirect link.

In spite of the preceding diatribe, it is useful to study the special properties of independent-risks systems. This is (a) to understand better the classical approach, (b) to disentangle properties which rely on independence from ones which do not, and (c) to identify the effects of a lack of independence. First, it is clear that when independent risks obtains the set of marginals $\{\bar{G}_j(t)\}$ determines the joint $\bar{G}(\boldsymbol{t})$ as $\Pi\bar{G}_j(t_j)$, and thereby all other probabilistic aspects of the set-up. For instance, Equation 3.1.1 yields

$$f(j, t) = g_j(t)\Pi_{k \neq j}\bar{G}_k(t) = \{g_j(t)/\bar{G}_j(t)\}\Pi_{k=1}^{p}\bar{G}_k(t) = h_j(t)\bar{F}(t) \quad (3.4.1)$$

in this case, where $g_j(t) = -d\bar{G}_j(t)/dt$ is the marginal density of T_j.

Example 3.7 Lagakos (1978)

Lagakos' model (Example 2.6) was based on independent exponential latent failure times for the two causes of death: he took $\bar{G}(\boldsymbol{t}) = \bar{G}_1(t_1)\bar{G}_2(t_2)$, with $\bar{G}_j(t) = e^{-\lambda_j t}$ ($j = 1, 2$). From Equation 3.4.1 then follows

$$f(j, t) = \lambda_j e^{-\lambda_+ t}, \ \bar{F}(j, t) = (\lambda_j/\lambda_+)e^{-\lambda_+ t},$$

where $\lambda_+ = \lambda_1 + \lambda_2$. ■

The following theorem gives some other implications of independent risks: in particular, it shows that the unobservable $\bar{G}_j(t)$ can then be derived explicitly from the observable $\bar{F}(j, t)$.

Theorem 3.3 (Gail, 1975)

The implications (i) ⇒ (ii) ⇒ (iii) ⇒ (iv) hold for the following propositions:

(i) independent risks obtains;
(ii) $h(j, t) = h_j(t)$ for each j and $t > 0$;
(iii) the set of sub-survivor functions $\bar{F}(j, t)$ determines the set of marginals $\bar{G}_j(t)$, explicitly,

$$G_j(t) = \exp\left\{-\int_0^t h(j, s)ds\right\}; \tag{3.4.2}$$

(iv) $\bar{G}(t\mathbf{1}_p) = \Pi_{j=1}^p \bar{G}_j(t)$.

Proof

Under (i), $\log\bar{G}(\boldsymbol{t}) = \sum_{j=1}^p \log\bar{G}_j(t_j)$ so, differentiating with respect to t_j and then setting $\boldsymbol{t} = t\mathbf{1}_p$,

$$[-\partial\log\bar{G}(\boldsymbol{t})/\partial t_j]_{t\mathbf{1}_p} = -d\log\bar{G}_j(t)/dt \tag{3.4.3}$$

for each j; this is just (ii). By Theorem 3.1, the left hand side of Equation 3.4.3 is equal to $f(j, t)/\bar{G}(t\mathbf{1}_p)$, which equals $f(j, t)/\Pi_{j=1}^p \bar{F}(j, t)$. Hence, $\bar{G}_j(t)$ on the right hand side is determined from the set of $\bar{F}(j, t)$ on the left hand side, and so (iii) holds. Equation 3.4.2 results from integrating Equation 3.4.3 with respect to t. From Equation 3.4.2,

$$\prod_{j=1}^p \bar{G}_j(t) = \exp\left\{-\sum_{j=1}^p \int_0^t h(j, s)ds\right\} = \exp\left\{-\int_0^t h(s)ds\right\} = \bar{F}(t),$$

which is (iv). ■

Part (iv) of the theorem looks a bit like independence of the T_j but isn't because that would require equality of $\bar{G}(\boldsymbol{t})$ and $\Pi\bar{G}_j(t_j)$ for all $\boldsymbol{t}$. What it does say is, in effect, that $\bar{R}(t\mathbf{1}_p) = 1$, where $\bar{R}(\boldsymbol{t})$ is the dependence measure defined in Section 3.1.1; this is "independence on the diagonal" $\boldsymbol{t} = t\mathbf{1}_p$.

3.4.1 *The Makeham assumption*

Condition (ii) of Theorem 3.3 is well known in the Competing Risks business. Gail (1975) called it the "Makeham assumption," after the 1874 paper, and we shall follow suit. Cornfield (1957) was also concerned with this assumption, and Elandt-Johnson (1976) referred to it as the "identity of forces of mortality." It says that the two hazard functions for risk j, the one in the presence of the other risks and the other in their hypothetical absence, are equal. Even when the risks are not independent, it can still hold for some j, as demonstrated explicitly by the example of Williams and Lagakos (1977, Section 3). That it can hold for all j when the risks are not independent is proved by Theorem 7.5 below: starting with an independent-risks model, and therefore with $h(j, t) = h_j(t)$ for all j, one can modify $\bar{G}(\boldsymbol{t})$ with infinite variety without disturbing any of the $h(j, t)$ or $h_j(t)$.

The Makeham assumption is closely related to the dependence measure $\bar{R}(t)$ via

$$[\partial \log \bar{R}(\mathbf{t})/\partial t_j]_{t\mathbf{1}_p} = h_j(t) - h(j, t).$$

A further implication is that

$$f(j, t) = \{g_j(t)/\bar{G}_j(t)\}\bar{F}(t) = g_j(t)\{\Pi_{k \neq j}\bar{G}_k(t)\}\bar{R}(t\mathbf{1}_p)$$
$$= g_j(t)\{\Pi_{k \neq j}\bar{G}_k(t)\}.$$

This says that $f(j, t)$ is deceptively expressible in the same form that it would have in the independent-risks case.

Under the Makeham assumption the likelihood contribution L_j in Equation 2.1.4 involves only the jth marginal survivor function $\bar{G}_j(t)$, remembering that $h_j(t) = -d\log \bar{G}_j(t)/dt$. However, even in this case, it is the $\bar{F}(j, t)$, not the $\bar{G}_j(t)$, that are of real-world relevance, as discussed in Section 3.3. In the special case of independence, the function $F(j, t)$ is given by

$$\int_t^\infty f(j, s)ds = \int_t^\infty g_j(s)\Pi_{k \neq j}\bar{G}_k(s)ds,$$

from Equation 3.4.1. Thus, even here, to make inferences about observable occurrences of failures of type j, one still needs to use all the $\bar{G}_j(t)$. The point is that observation of a failure of type j necessarily involves the prior non-occurrence of failures of all other types.

Part (iii) of Theorem 3.3 says that, under independent risks, the set of observable distributions $\bar{F}(j, t)$ and the set of marginal distributions $\bar{G}_j(t)$ each give complete descriptions of the set-up. That (i) $\Rightarrow$ (iii) has been proved by Berman (1963) and Nadas (1970); these papers also give Formula 3.4.2 in slightly different forms.

3.4.2 Proportional hazards

For the case of independent risks, the equivalence of proportional hazards and independence of time and cause of failure (Theorem 1.1) has been shown in various versions by Allen (1963), Sethuraman (1965), and Nadas (1970a).

The following result has been given for the special case of independent risks by Armitage (1959) and Allen (1963). As now shown, it actually holds under the weaker, Makeham assumption.

Lemma 3.4 (Crowder, 1993)

Under the Makeham assumption, proportional hazards obtains if and only if $\bar{G}_j(t) = \bar{F}(t)^{p_j}$.

Proof

Proportional hazards means $h(j, t) = p_j h(t)$. Under the Makeham assumption, that $h(j, t) = h_j(t)$, this is $h_j(t) = p_j h(t)$, i.e.,

$$d\log\bar{G}_j(t)/dt = p_j d\log\bar{F}(t)/dt .$$

On integration, this gives the result. ■

The relation in the lemma, $\bar{G}_j(t) = \bar{F}(t)^{p_j}$, is reminiscent of that in Theorem 1.1, $\bar{F}(j, t) = p_j\bar{F}(t)$. However, that in the theorem is rather more fundamental in the sense that it is equivalent to proportional hazards whether or not the Makeham assumption holds, in fact, whether or not there are any latent failure times associated with the set-up at all.

A consequence of the lemma is that in the independent-risks case $\bar{G}(\boldsymbol{t})$ can be expressed as $\Pi_{j=1}^{p}\bar{F}(t_j)^{p_j}$, i.e., solely in terms of the function $\bar{F}(t)$. For statistical modelling this means that we only have one univariate survivor function $\bar{F}(t)$ to worry about, rather than a whole set of $\bar{F}(j, t)$ or a multivariate $\bar{G}(\boldsymbol{t})$. Suppose, for example, that each $\bar{F}(j, t)$ requires q parameters when modelled separately. Then the full set would require pq parameters that would be reduced to $q + p - 1$ under proportional hazards, the extra $p - 1$ parameters being the p_js.

David (1970) noted that the relation $\bar{F}(t) = \bar{G}_j(t)^{1/p_j}$ has the same form as one which represents the survivor function of the minimum of an integer number, $1/p_j$, of independent variates each with survivor function $\bar{G}_j(t)$. This led him to suggest the use of extreme value distributions, in particular, Weibull with $\bar{F}(t) = \exp\{-(t/\xi)^{\phi}\}$. In a later article David (1974) reviewed some basic theory and considered parametric maximum likelihood estimation, in particular for independent Weibull latent failure times. He also discussed the Marshall-Olkin system (Section 7.2) and gave a generalization.

3.4.3 Some examples

Gasemyr and Natvig (1994) have outlined a Bayesian approach to estimating component latent lifetimes based on autopsy data.

Example 3.8

Berkson and Elveback (1960) considered two independent risks with exponential failure times. They gave some basic formulae, discussed estimation, including maximum likelihood, and applied the methods to a prospective study relating smoking habits to time and cause of death among 200,000 U.S. males. Their model can be written in terms of the sub-distribution functions as

$$F(j, t) = (\alpha_j/\alpha_+)\{1 - e^{-\alpha_+ t}\},$$

for non-smokers, and

$$F(j, t) = (\beta_j/\beta_+)\{1 - e^{-\beta_+ t}\},$$

for smokers; $j = 1$ for death from lung cancer, $j = 2$ for death from other causes, $\alpha_+ = \alpha_1 + \alpha_2$ and $\beta_+ = \beta_1 + \beta_2$. The comparison of particular interest is between β_j and α_j.

Example 3.9

Herman and Patel (1971) assumed independent risks, for which they summarized the basic probability functions. They went on to consider, for two risks, parametric maximum likelihood estimation for Type 1 (fixed time) censored samples and specialized their formulae to exponential and Weibull latent failure times. They re-analyzed Mendenhall and Hader's (1958) data based on the exponential model.

Example 3.10

Klein and Basu (1981) considered "Weibull accelerated life tests." Their model is one of independent risks with marginal hazard functions

$$h_j(t;x) = \psi_{jx} h_0(t),$$

where $h_0(t) = \phi_j t^{\phi_j - 1}$ has the Weibull hazard form with shape parameter ϕ_j, the scale parameter being absorbed in $\psi_{jx} = \exp(x^T\beta_j)$ which combines the vector x of explanatory variables with the regression coefficients β_j. At first sight, this looks more like a univariate proportional hazards specification than an accelerated life one, but the two coincide for the Weibull distribution. The authors computed likelihood functions and derivatives explicitly for Type 1 (fixed time), Type 2 (fixed number of failures), and progressively (hybrid scheme) censored samples.

3.5 A risk-removal model

Hoel (1972) developed a model which allows individuals in the population to have their own immunity profiles. Each individual has a set of risk indicators $(r_1, \ldots, r_p)$ such that $r_j = 1$ if he is vulnerable to risk j and $r_j = 0$ if not, the proportion vulnerable being q_j in the population. Technically, the latent failure time T_j is represented as S_j/r_j, where the S_j and r_j are all independent, S_j has distribution function $G_{s_j}(t)$, and $\mathrm{pr}(r_j = 1) = q_j$. As usual, $T = \min\{T_1, \ldots, T_p\} = T_C$. In effect, the T_j are being turned into improper random variables in the sense that $\mathrm{pr}(T_j > t) \to 1 - q_j$, not 0, as $t \to \infty$. Hoel took one of the q_j to be equal to 1 to make T finite with probability 1 — no immortality here, then. The r_j are not directly observed: they are random effects which vary from case to case. If the r_j were known, one could just condition out the T_j for which $r_j = 0$, as suggested by Elandt-Johnson (1976, see Section 3.2 above). The q_j are parameters of the model.

The marginal survivor function of T_j can be obtained as

$$\begin{aligned}\bar{G}_j(t) &= \mathrm{pr}(T_j > t) = \mathrm{pr}(S_j > tr_j) \\ &= \mathrm{pr}(S_j > t)q_j + \mathrm{pr}(S_j > 0)(1 - q_j) = 1 - q_j G_{s_j}(t).\end{aligned} \tag{3.5.1}$$

Hence,

$$\bar{F}(t) = \Pi_{j=1}^p \bar{G}_j(t) = \Pi_{j=1}^p \{1 - q_j G_{s_j}(t)\},$$

$$h(j, t) = h_j(t) = g_j(t)/\bar{G}_j(t) = q_j g_{s_j}(t)/\{1 - q_j G_{s_j}(t)\},$$

where $g_{s_j}(t) = dG_{s_j}(t)/dt$ is the density of S_j, and $f(j, t) = h(j, t)\bar{F}(t)$.

Hoel applied his model to two groups of mortality data, each with three risks; we borrowed his data earlier for Example 2.5. For $G_{s_j}(t)$, he used the form $1 - \exp\{-\phi_j(1 - e^{\lambda_j t})\}$. There were no explanatory variables, other than the grouping, and he employed parametric maximum likelihood estimation and used some goodness-of-fit indicators and plots (Section 2.2).

A dependent-risks version of Hoel's model can be derived as follows. We have

$$\bar{G}(\boldsymbol{t}) = \mathrm{pr}\{\cap_{j=1}^p (S_j > r_j t_j)\} = \sum_{\boldsymbol{r}} g(\boldsymbol{r})\bar{G}_s(r_1 t_1, \ldots, r_p t_p), \tag{3.5.2}$$

where the summation $\sum_{\boldsymbol{r}}$ is over all binary vectors $\boldsymbol{r} = (r_1, \ldots, r_p)$, $\bar{G}_s(\boldsymbol{s})$ is the joint survivor function of $\boldsymbol{S} = (S_1, \ldots, S_p)$, and $g(\boldsymbol{r})$ is the joint probability function of $\boldsymbol{r}$ given by

$$g(\boldsymbol{r}) = \Pi_{j=1}^p \{q_j^{r_j}(1 - q_j)^{1 - r_j}\}.$$

For example, for $p = 2$,

$$\begin{aligned}\bar{G}(\boldsymbol{t}) &= \bar{G}_s(t_1, t_2)q_1 q_2 + \bar{G}_s(0, t_2)(1 - q_1)q_2 \\ &\quad + \bar{G}_s(t_1, 0)q_1(1 - q_2) + \bar{G}_s(0, 0)(1 - q_1)(1 - q_2),\end{aligned}$$

and for independent S_j Equation 3.5.2 reduces to $\Pi_{j=1}^p \{1 - q_j G_{s_j}(t_j)\}$ in accordance with Equation 3.5.1. The associated probability functions are, from Equation 3.5.2,

$$\bar{F}(t) = \bar{G}(t\boldsymbol{1}_p) = \sum_{\boldsymbol{r}} g(\boldsymbol{r})\bar{G}_s(t\boldsymbol{r})$$

and, using Equation 3.1.1,

$$f(j, t) = [-\partial \bar{G}(\boldsymbol{t})/\partial t_j]_{t1_p} = \sum_{\boldsymbol{r}} g(\boldsymbol{r}) r_j [-\partial \bar{G}_s(\boldsymbol{s})/\partial s_j]_{t\boldsymbol{r}}.$$

chapter 4

Likelihood functions for univariate survival data

We take the day off from Competing Risks now in order to do some spadework. The main topic is the semi-parametric methods that have come to the fore in recent years. In this chapter we give an overview of their application in Survival Analysis and develop this for Competing Risks in the next. The chapter starts with discrete failure times, goes on to continuous ones in the middle, then ends up with discrete ones again. Don't ask — it just provides a more logical way to develop the methodology.

4.1 Discrete and continuous failure times

At first sight, discrete failure times might seem to be of limited application in Survival Analysis, but when one considers more carefully the definition of "failure time," it is seen not to be so limited in practice. Thus, failure of a system can be defined in the obvious sense of blow-up or breakdown, or, more generally, by an assessment that it is no longer able to perform its function satisfactorily, e.g., from observable degradation. Such assessments may be made automatically or by human inspection.

Consider a system in continuous operation that is inspected periodically. Suppose that if one or more components are classified as effectively "life-expired," the system is taken out of service for renewal or replacement. Then, discounting the possibility of breakdown between inspections, the lifetime of a component is the number of inspections to life-expiration. A similar situation occurs when a continuously operating system is subject to shocks, peak stresses, loads, or demands. Such stresses can be regular, such as those caused by tides or weekly demand cycles, or irregular, as with storms and earthquakes. The lifetime for such a system is the number of shocks until failure.

Some systems operate intermittently rather than continuously. The cycles of operation may be regular or irregular, and the system lifetime is the number of cycles until failure. Regular operation is commonly encountered in such areas as manufacturing production runs, cycles of an electrical system, machines producing individual items, and orbiting satellites exposed to solar radiation each time they emerge from the earth's shadow. Another example occurs where certain electrical and structural units on aircraft have to be inspected after each flight: the location of one or more critical faults will lead to "failing" the system. There are also many examples of irregular operation. Some systems have to respond to unpredictable demands, a typical task for standby units. In this section some of the basics are set out.

4.1.1 Discrete survival distributions

Let T be a discrete failure time taking possible values $0 = \tau_0 < \tau_1 < \ldots < \tau_m$; m may be finite or infinite, as may τ_m. The survivor function is $\bar{F}(t) = \text{pr}(T > t)$, as usual, and the corresponding discrete density function is defined as $f(t) = \text{pr}(T = t)$. They are related by $f(t) = \bar{F}(t-) - \bar{F}(t)$ and $\bar{F}(t) = \sum f(\tau_s)$, where the summation is over $\{s:\tau_s > t\}$. The notation $\bar{F}(t-)$ is a useful abbreviation for $\lim_{\delta \to 0} \bar{F}(t - \delta)$, where $\delta > 0$. (For continuous failure times, $\bar{F}(t-) = \bar{F}(t)$.) If t is not equal to one of the τ_l, $f(t) = 0$. Also, we will adopt the convention $\bar{F}(\tau_0) = 1$, i.e., $f(0) = 0$, so that zero lifetimes are ruled out. The discrete hazard function is defined as

$$h(t) = \text{pr}(T = t | T \geq t) = f(t)/\bar{F}(t-).$$

(For continuous failure times, the denominator is usually written equivalently as $\bar{F}(t)$.) Note that $0 \leq h(t) \leq 1$ for all t, with $h(t) = 0$ except at the points τ_l ($l = 1, \ldots, m$), $h(0) = 0$, and $h(t) = 1$ only at the upper end point τ_m of the distribution, i.e., where $\bar{F}(t-) = f(t)$. The inverse relationship, expressing $\bar{F}(t)$ in terms of $h(t)$, can be derived as

$$\bar{F}(t) = \bar{F}(t-) - f(t) = \bar{F}(t-)\{1 - h(t)\} = \Pi_{s=1}^{l(t)}\{1 - h(\tau_s)\},$$

where $l(t) = \max\{l:\tau_l \leq t\}$ so that the product is over $\{s:\tau_s \leq t\}$. Also,

$$f(t) = h(t)\Pi_{s=1}^{l(t-)}\{1 - h(\tau_s)\}.$$

Otherwise expressed, and writing h_l for $h(\tau_l)$, these representations are

$$\bar{F}(\tau_l) = \Pi_{s=1}^{l}(1 - h_s) \text{ for } l \geq 1, \quad f(\tau_l) = h_l\Pi_{s=1}^{l-1}(1 - h_s) \text{ for } l \geq 2, \tag{4.1.1}$$

with $\bar{F}(\tau_0) = 1$, $\bar{F}(\tau_m) = 0$, $f(\tau_0) = 0 = h_0$, and $f(\tau_1) = h_1$. The integrated hazard function, defined in Equation 1.1.1, is

$$H(t) = -\log\bar{F}(t) = -\sum_s \log(1 - h_s).$$

If the h_s in the summation are small, then $\log(1 - h_s) \approx -h_s$ and $H(t) \approx \sum h_s$, which can justifiably be called the **cumulative hazard function**.

Example 4.1 Geometric distribution

This is just about the simplest discrete waiting time distribution. It arises as the time to failure in a sequence of Bernoulli trials, i.e., independent trials each with probability ρ of success. In the present context the process carries on while it's still winning. The τ_l here are the non-negative integers: $\tau_l = l$ for $l = 0, 1, \ldots, m = \infty$. We have $f(0) = 0$ and, for $l \geq 1$,

$$f(l) = \rho^{l-1}(1-\rho),\ \bar{F}(l) = \rho^{l},\ h_l = (1-\rho).$$

Note that the hazard function is constant, independent of l, as for the exponential among continuous distributions. Further, the famed **lack of memory** property of the exponential is also shared by the geometric:

$$\mathrm{pr}(T > t + s | T > s) = \bar{F}(t+s)/\bar{F}(s) = \rho^{t} = \bar{F}(t) = \mathrm{pr}(T > t).$$

Actually, of course, this is really only another way of saying that the hazard function is constant, as can be seen by considering the general identity

$$\bar{F}(t+s)/\bar{F}(s) = \Pi_{r=s+1}^{s+t}(1-h_r):$$

if h_r is independent of r this expression is equal to $\bar{F}(t)$; conversely, if the expression is equal to $\bar{F}(1)$ for $t = 1$ and every s, then $\bar{F}(1) = 1 - h_{s+1}$ and so h_{s+1} is independent of s. ■

4.1.2 *Mixed survival distributions*

Suppose that T has a mixed discrete/continuous distribution, with atoms of probability at points $0 = \tau_0 < \tau_1 < \tau_2 < \ldots$, together with a density $f^c(t)$ on $(0, \infty)$. If $\bar{F}$ is continuous at t, $\bar{F}(t-) = \bar{F}(t)$, whereas if $t = \tau_l$, $\bar{F}(\tau_l-) = \bar{F}(\tau_l) + \mathrm{pr}(T = \tau_l)$. Then

$$\bar{F}(t) = \mathrm{pr}(T > t) = \sum_{\tau_l > t} \mathrm{pr}(T = \tau_l) + \int_t^{\infty} f^c(s)ds;$$

$f^c(t) = -d\bar{F}(t)/dt$ is defined at points between the τ_l. Now,

$$\bar{F}(t) = \bar{F}(0)\Pi_{l=1}^{l(t)}[\{\bar{F}(\tau_1)/\bar{F}(\tau_l-)\}\{\bar{F}(\tau_l-)/\bar{F}(\tau_{l-1})\}]\{\bar{F}(t)/\bar{F}(\tau_{l(t)})\} \quad (4.1.2)$$

where $l(t) = \max\{l : \tau_l \leq t\}$. But $\bar{F}(0) = 1 - \mathrm{h}_0$, where $h_0 = \mathrm{pr}(T = 0)$, and

$$\bar{F}(\tau_l)/\bar{F}(\tau_l-) = 1-\mathrm{pr}(T=\tau_l)/\bar{F}(\tau_l-) = 1-\mathrm{pr}(T\le\tau_l|T>\tau_l-) = 1-h_l,$$

say. Also,

$$\bar{F}(\tau_l-)/\bar{F}(\tau_{l-1}) = \exp\left\{-\int_{\tau_{l-1}}^{\tau_l-} h^c(s)ds\right\},$$

where

$$h^c(s) = f^c(s)/\bar{F}(s) = -d\log\bar{F}(s)/ds.$$

The h_l are the discrete hazard contributions at the discontinuities, and h^c is the continuous component of the hazard function. Last,

$$\bar{F}(t)/\bar{F}(\tau_{l(t)}) = \exp\left\{-\int_{\tau_{l(t)}}^{t} h^c(s)ds\right\},$$

which equals 1 if $t = \tau_{l(t)}$. Substituting into Equation 4.1.2 yields the well-known formula (e.g., Cox, 1972, Section 1):

$$\bar{F}(t) = \{\Pi_{s=1}^{l(t)}(1-h_s)\}\exp\left\{-\int_0^t h^c(s)ds\right\}. \tag{4.1.3}$$

Consider now a purely discrete distribution in which the density component is absent, so that $h_l = \mathrm{pr}(\tau_{l-1} < T \le \tau_l \mid T > \tau_{l-1})$. Let $\delta_l = \tau_l - \tau_{l-1}$ and $g(\tau_l) = h_l/\delta_l$, so that, in the limit $\delta_l \to 0$, $g(\tau_l)$ is defined as the hazard function at τ_l of a continuous variate. Now

$$\log\bar{F}(t) = \log\prod_{\tau_l\le t}(1-h_l) = \sum_{\tau_l\le t}\log\{1-g(\tau_l)\delta_l\}$$

$$= -\sum_{\tau_l\le t} g(\tau_l)\delta_l + O\left\{\sum_{\tau_l\le t} g(\tau_l)^2\delta_l^2\right\}.$$

In the limit max $(\delta_l) \to 0$ we obtain

$$\bar{F}(t) = \exp\left\{-\int_0^t g(s)ds\right\}.$$

The transition from an increasingly dense discrete distribution to a continuous one is thus illustrated. This is just an informal sketch of material dealt with in much greater depth by Gill and Johansen (1990, Section 4.1).

The reverse transition, obtained by dividing up the continuous time scale into intervals (τ_{l-1}, τ_l), is accomplished simply by defining

$$h_l = 1 - \exp\left\{-\int_{\tau_{l-1}}^{\tau_l} g(s)ds\right\}.$$

Then

$$\bar{F}(\tau_k) = \exp\left\{-\int_0^{\tau_k} g(s)ds\right\} = \exp\left\{-\sum_{l=1}^k \int_{\tau_{l-1}}^{\tau_l} g(s)ds\right\} = \prod_{s=1}^k (1-h_l).$$

Example 4.2 Geometric and exponential distributions

A standard textbook connection between the two distributions is as follows. Suppose that the continuous time scale is divided into equal intervals of length δ, and that independent Bernoulli trials are performed with probability $1 - \pi = \lambda\delta$ of failure within each interval. Let $M = T/\delta$, the number of intervals survived without failure. Then M has the geometric survivor function $\text{pr}(M > m) = \pi^m$ for $m = 1, 2, \ldots$; hence, $\text{pr}(T/\delta > t/\delta) = (1 - \lambda\delta)^{t/\delta}$. As $\delta \to 0$, $\text{pr}(T > t) \to e^{-\lambda t}$, i.e., an exponential distribution. A slightly modified version can be based on the discrete survivor function $\text{pr}(M > m) = \pi^{m^\phi}$, with $\phi > 0$; π is the probability of success on the first trial, and $\pi^{m^\phi - (m-1)^\phi}$ is the probability of success on the mth given survival that far. This leads to $(1 - \lambda\delta^\phi)^{(t/\delta)^\phi}$ and thence, allowing $\delta \to 0$, to a Weibull distribution with $\text{pr}(T > t) = \exp(-\lambda t^\phi)$. ■

Example 4.3 Negative binomial and gamma distributions

The negative binomial is often introduced as a waiting time distribution: let M be the number of Bernoulli trials performed ending with the kth failure, then

$$\text{pr}(M = k + m) = \binom{k+m-1}{k-1}\pi^k(1-\pi)^m \qquad (m = 0, 1, \ldots).$$

Obviously, M can be represented as the sum of k consecutive waiting times to a first failure, i.e., as the sum of k independent geometric variates. Then the limiting process described in the preceding example yields the sum of k independent exponential variates with rate parameter λ, i.e., a gamma distribution for T with density $\lambda^k t^{k-1} e^{-\lambda t}/\Gamma(k)$. ■

4.2 Discrete failure times: estimation

We will consider the construction of a likelihood function for the discrete case since this shows the basic approach, in terms of hazard functions, for the other likelihoods in this chapter.

4.2.1 Random samples: parametric estimation

We assume for the moment that there are no explanatory variables. It is understood that a case classified as being right-censored at τ_l was observed not to fail at τ_l. Suppose that the sample comprises r_l observed failures and s_l right-censored ones at time τ_l ($l = 0, \ldots, m$); in accordance with the definitions of τ_0 and τ_m, $r_0 = 0$, $r_m \geq 0$, $s_0 \geq 0$, and $s_m = 0$. Obviously, with a finite sample, the recorded values of r_l and s_l will all be zero for l beyond some finite point. The likelihood contributions are, respectively, $f(\tau_l)$ for an observed failure time τ_l and $\bar{F}(\tau_l)$ for one right-censored at τ_l. The overall likelihood function is then

$$L = \Pi_{l=1}^{m}\{f(\tau_l)^{r_l}\bar{F}(\tau_l)^{s_l}\}, \tag{4.2.1}$$

where $\bar{F}(\tau_m)^{s_m}$ is interpreted as 1; there is no explicit factor for $l = 0$ because $r_0 = 0$ and $\bar{F}(\tau_0) = 1$. This likelihood can be used for parametric inference in the usual way after substituting the chosen parametric forms for f and $\bar{F}$. Note that L does not involve s_0, reflecting the absence of information in lifetimes known only to be positive.

The insertion of $\bar{F}(\tau_l)$ in Equation 4.2.1 assumes that the censoring tells us nothing about the failure time except that it is beyond τ_l. It is not always the case that censoring is non-informative, e.g., in some circumstances censoring is associated with imminent failure. The topic will be covered more fully in Section 7.6.

Example 4.1 (continued)

For this distribution, Equation 4.2.1 takes the form, after setting $\tau_l = l$,

$$L(\rho) = \Pi_{l=1}^{m}[\{\rho^{l-1}(1-\rho)\}^{r_1}(\rho^l)^{s_l}] = \rho^a(1-\rho)^b,$$

where $a = \Sigma\{ls_l + (l-1)r_l\}$ and $b = \Sigma r_l$. Thus, the maximum likelihood estimator is $\hat{\rho} = a/(a+b)$. The second derivative of the log-likelihood is

$$L''(\rho) = -a/\rho^2 - b/(1-\rho)^2$$

and so the asymptotic variance of $\hat{\rho}$ can be estimated by

$$\{-L''(\rho)\}^{-1} = ab/(a+b)^3.$$

To calculate the expected information $E\{-L''(\rho)\}$ we would need to know more about the censoring rule which determines the distribution of the r_l and s_l. Suppose, for example, that the sample size $n = r_+ + s_+$ is fixed, and that everything is observed up to time m', so that $r_l = 0$ for $l > m'$ and $s_l = 0$ for $l \neq m'$. Then $(r_1, \ldots, r_{m'}, s_{m'})$ has a multinomial distribution with probabilities $\pi_l = \rho^{l-1}(1-\rho)$ for $l = 1, \ldots, m'$ and $\pi_{m'+1} = \rho^{m'+1}$. Hence,

$$\begin{aligned} E(a) &= \sum_{l=1}^{m'} lE(s_l) + \sum_{l=1}^{m'} (l-1)E(r_l) = nm'\rho^{m'+1} + n\sum_{l=2}^{m'} (l-1)\rho^{l-1}(1-\rho) \\ &= nm'\rho^{m'+1} - nm'\rho^{m'} + n\rho(1-\rho^{m'})/(1-\rho), \\ E(b) &= \sum_{l=1}^{m'} E(r_l) = \sum_{l=1}^{m'} n\rho^{l-1}(1-\rho) = n(1-\rho^{m'}) \end{aligned}$$

which gives $E\{-L''(\rho)\}$ explicitly in terms of n, m', and ρ. ■

4.2.2 Random samples: non-parametric estimation

For nonparametric estimation we use the hazard representation, Equation 4.1.1. Thus, writing h_l for $h(\tau_l)$, Equation 4.2.1 becomes

$$L = \Pi_{l=1}^{m}[\{h_1\Pi_{s=0}^{l-1}(1-h_s)\}^{r_l} \times \{\Pi_{s=0}^{l}(1-h_s)\}^{s_1}].$$

After collecting terms,

$$L = \Pi_{l=1}^{m}\{h_l^{r_l}(1-h_l)^{q_l-r_l}\}, \tag{4.2.2}$$

where

$$q_l = (r_l + s_l) + (r_{l+1} + s_{l+1}) + \ldots + (r_m + s_m);$$

q_l is the number of cases still **at risk** at time τ_l-, q_0 being the overall sample size. Here, "at risk" means still at risk of being recorded as a failure, i.e., cases which have not yet failed, nor been lost to view by (non-informative) censoring, and are therefore still under observation. Treating the h_l as the parameters of the distribution, their maximum likelihood estimators are $\hat{h}_l = r_l/q_l$ ($l = 1, \ldots, m$), obtained by solving $\partial \log L/\partial h_l = 0$. The $\hat{h}_l$ evidently result from equating h_l, the conditional probability of failure at time τ_l, to r_l/q_l, the proportion of observed failures among those liable to observed failure at that time. The corresponding estimators for $\bar{F}$ and f follow from Equation 4.1.1:

$$\hat{\bar{F}}(\tau_l) = \Pi_{s=1}^{l}(1-\hat{h}_s) \text{ for } l \geq 1, \ \hat{f}(\tau_1) = \hat{h}_l\Pi_{s=1}^{l-1}(1-\hat{h}_s) \text{ for } l \geq 2, \tag{4.2.3}$$

and $\hat{f}(\tau_l) = \hat{h}_l$.

Note that, since $r_0 = 0$, the formula $\hat{h}_l = r_l/q_l$ extends correctly to cover $l = 0$, and then $\hat{\bar{F}}(\tau_0) = 1 - \hat{h}_0 = 1$. Likewise, since $s_m = 0$ necessarily, $\hat{h}_m = r_m/q_m = 1$ (interpreting 0/0 as 1, if necessary) and then $\hat{\bar{F}}(\tau_m) = 0$ correctly. In principle, s_{m-1} could be transferred to r_m, since individuals observed to survive beyond time τ_{m-1} must then fail at time τ_m. However, this makes no difference to the estimates $\hat{h}_{m-1}$ and $\hat{h}_m$.

In finite samples it is possible for q_l to become zero first at $l = m' \le m$ and so $q_l = 0$ for $l = m', \ldots, m$; this is bound to occur if $m = \infty$. The estimates $\hat{h}_l = r_l/q_l$, as 0/0, are then formally undefined for $l \ge m'$. This lack of an estimate for h_l $(l \ge m')$ can be put down to the fact that the conditioning event $\{T > \tau_{l-1}\}$, essential to its definition, has not been observed. However, if $s_{m'-1} = 0$, then $q_{m'-1} = r_{m'-1}$ and $\hat{h}_{m'-1} = 1$, which gives $\hat{\bar{F}}(\tau_{m'-1}) = 0$ and thence $\hat{\bar{F}}(\tau_l) = 0$ for $l = m', \ldots, m$. If $s_{m'-1} > 0$, we know only that

$$\hat{\bar{F}}(\tau_{m'-1}) \ge \hat{\bar{F}}(\tau_l) \ge \hat{\bar{F}}(\tau_{l+1}) \ge 0 \quad \text{for } l = m', \ldots, m-1,$$

with $\hat{\bar{F}}(\tau_m) = 0$.

In the absence of censoring, $s_l = 0$ for all l, and then $1 - \hat{h}_l = q_{l+1}/q_l$ and $\hat{\bar{F}}(\tau_l) = q_{l+1}/q_1$, the usual **empirical survivor function**, noting that $q_1 = q_0$ in this case.

We now restrict attention to the case $m < \infty$. Then there is a finite number of parameters h_l and, under standard regularity conditions, the usual large-sample parametric theory will apply. Thus, the asymptotic joint distribution of the set of $(\hat{h}_l - h_l)$ will be multivariate normal with means zero and covariance matrix $\mathrm{E}(\boldsymbol{H})^{-1}$, where $\boldsymbol{H}$ has (l, k) element

$$H_{lk} = -\partial^2 \log L/\partial h_l \partial h_k = \delta_{lk}\{r_l/h_l^2 + (q_l - r_l)/(1 - h_l)^2\};$$

here δ_{lk} is the Kronecker delta taking value 1 if $l = k$ and 0 if $l \ne k$. The diagonal form of $\boldsymbol{H}$, implying asymptotic independence of the $\hat{h}_l$, results from the orthogonal form of L (Cox and Reid, 1987). Using the observed, rather than the expected, information (Section 2.2.2), we have the simpler form

$$\hat{H}_{lk} = \delta_{lk}\{q_l^2/r_l + q_l^2/(q_l - r_l)\} = \delta_{lk} q_l/\{\hat{h}_l(1 - \hat{h}_l)\},$$

and so $\mathrm{var}(\hat{h}_l)$ can be estimated by $\hat{h}_l(1 - \hat{h}_l)/q_l$. Applying the delta method (e.g., Crowder et al., 1991, Appendix) to Equation 4.2.3 we have

$$\mathrm{var}\{\log \hat{\bar{F}}(t)\} = \mathrm{var}\left\{\sum_{\tau_s \le t} \log(1 - \hat{h}_s)\right\} \approx \sum (1 - h_s)^{-2} \mathrm{var}(1 - \hat{h}_s)$$

$$\approx \sum (1 - \hat{h}_s)^{-2} \hat{h}_s(1 - \hat{h}_s)/q_s.$$

Thus, we obtain **Greenwood's formula**, so named after his 1926 contribution:

$$\mathrm{var}\{\hat{\bar{F}}(t)\} \approx \hat{\bar{F}}(t)^2 \sum \hat{h}_s/\{(1-\hat{h}_s)q_s\}. \tag{4.2.4}$$

Actually, the formula for $\mathrm{var}\{\log\hat{\bar{F}}(t)\}$ is probably more useful than Equation 4.2.4 because $\log\bar{F}(t)$ has less restricted range than $\bar{F}(t)$. The alternative transforms $\log\{-\log\bar{F}(t)\}$ and $\log[\bar{F}(t)/\{1-\bar{F}(t)\}]$ are fully unrestricted, and some authors have recommended working in terms of these if variances are to be used in this context (e.g., Kalbfleisch and Prentice, 1980, Section 1.3). If there is no censoring,

$$\hat{h}_s/\{(1-\hat{h}_s)q_s\} = r_s/\{(q_s-r_s)q_s\} = (q_s-q_{s+1})/(q_{s+1}q_s) = q_{s+1}^{-1}-q_s^{-1}.$$

Then, writing $\hat{\bar{F}}_l$ for $\hat{\bar{F}}(\tau_l) = q_{l+1}/q_1$, Equation 4.2.4 becomes

$$\hat{\bar{F}}_l^2(q_l^{-1}-q_1^{-1}) = \hat{\bar{F}}_l(1-\hat{\bar{F}}_l)/q_1,$$

the usual binomial variance estimate.

4.2.3 *Explanatory variables*

Suppose now that there are explanatory variables, or covariates, so that the *i*th individual comes complete with a vector x_i of such at no extra charge. Write $f(t;x_i)$ and $\bar{F}(t;x_i)$, respectively, for the density and survivor function of the failure time of the *i*th individual. The likelihood function Equation 4.2.1 is now filled out to

$$L = \Pi_{l=1}^{m}\{\Pi_{i\in R_l} f(\tau_l;x_i) \times \Pi_{i\in S_l}\bar{F}(\tau_l;x_i)\}, \tag{4.2.5}$$

where R_l is the set of individuals observed to fail at time τ_l, and S_l is the set of individuals right-censored at time τ_l. We take R_0 as empty, and note that individuals in S_0 turned up to sign on at time $\tau_0 = 0$, but had disappeared before the first roll call at time τ_1; likewise, R_m is the set of diehards who held out until time τ_m, and S_m is empty. (In principle, the set of individuals in S_{m-1} could be transferred to R_m since, although lost sight of just after time τ_{m-1}, they must fail at time τ_m.) The hazard contributions will be denoted by

$$h_l(x) = \mathrm{pr}(\text{failure at time } \tau_l \mid \text{survival past } \tau_{l-1};x) = f(\tau_l;x)/\bar{F}(\tau_{l-1};x),$$

and so Equation 4.1.1 gives:

$$\bar{F}(\tau_l;x) = \Pi_{s=1}^{l}\{1-h_s(x)\} \text{ for } l \geq 1,$$

$$f(\tau_l;x) = h_l(x)\Pi_{s=1}^{l-1}\{1-h_s(x)\} \text{ for } l \geq 2,$$

with $f(\tau_1;x) = h_1(x)$. Thus, the likelihood Equation 4.2.5 can be expressed in terms of the hazards as

$$\begin{aligned} L &= \Pi_{l=1}^{m}\{\Pi_{i\in R_l}[h_l(\boldsymbol{x}_i)\Pi_{s=1}^{l-1}\{1-h_s(\boldsymbol{x}_i)\}]\times\Pi_{i\in S_l}[\Pi_{s=1}^{l}\{1-h_s(\boldsymbol{x}_i)\}]\} \\ &= \Pi_{l=1}^{m}\{\Pi_{i\in R_l}h_l(\boldsymbol{x}_i)\times\Pi_{i\in R_l}\Pi_{s=1}^{l-1}g_{si}\times\Pi_{i\in S_l}\Pi_{s=1}^{l}g_{si}\}, \end{aligned} \tag{4.2.6}$$

where we have written g_{si} for $1\text{-}h_s(x_i)$ and $\Pi_{s=1}^{l-1}g_{si}$ is interpreted as 1 for $l = 1$. But, the terms involving g_{si} can be collected as

$$\Pi_{i\in R_2}g_{1i}\times\Pi_{i\in R_3}(g_{1i}g_{2i})\times\ldots\times\Pi_{i\in S_1}g_{1i}\times\Pi_{i\in S_2}(g_{1i}g_{2i})\times\ldots = \Pi_{l=1}^{m}\Pi_{i\in Q_l}g_{li},$$

where

$$Q_l = S_l \cup R_{l+1} \cup S_{l+1} \cup \ldots \cup R_m \cup S_m = R(t_l) - R_l, \tag{4.2.7}$$

and $R(t_l)$ is the **risk set** at time t_l, i.e., the set of individuals still at risk at time t_l- of eventually being recorded as failures. Thus, the likelihood function becomes

$$\begin{aligned} L &= \Pi_{l=1}^{m}[\Pi_{i\in R_l}h_l(\boldsymbol{x}_i)\times\Pi_{i\in Q_l}\{1-h_l(\boldsymbol{x}_i)\}] \\ &= \Pi_{l=1}^{m}\{\Pi_{i\in R_l}[h_l(\boldsymbol{x}_i)/\{1-h_l(\boldsymbol{x}_i)\}]\times\Pi_{i\in R(t_l)}[1-h_l(\boldsymbol{x}_l)]\}. \end{aligned} \tag{4.2.8}$$

This likelihood can be used for parametric or semi-parametric inference. First, we give a simple form of the former, and do the latter later.

Example 4.1 (continued)

Assume a logit model for $\rho_i = \rho(x_i)$: $\log\{\rho_i/(1-\rho_i)\} = \boldsymbol{x}_i^{\mathrm{T}}\beta$. Then,

$$h_l(\boldsymbol{x}_i) = 1-\rho_i = (1+e^{\boldsymbol{x}_i^{\mathrm{T}}\beta})^{-1},$$

and the likelihood Equation 4.2.8 for the regression parameter vector β is

$$L(\beta) = \Pi_{l=1}^{\infty}\left\{\Pi_{i\in R_l}e^{-\boldsymbol{x}_i^{\mathrm{T}}\beta} \div \Pi_{i\in R(t_l)}\left(1+e^{-\boldsymbol{x}_i^{\mathrm{T}}\beta}\right)\right\}. \blacksquare$$

4.2.4 *Interval-censored data*

The analysis outlined above, nominally for discrete failure times, actually applies without much damage to interval-censored, or grouped, continuous failure times of the kind recorded in life tables.

Suppose that the time scale is partitioned into intervals $(\tau_{l-1}, \tau_l]$ ($l = 1, \ldots, m$), with $\tau_0 = 0$ (and $\tau_{m+1} = \infty$). The data comprises, for each l, a set R_l of r_l individuals with failures at unknown times during $(\tau_{l-1}, \tau_l]$, and a set S_l of s_l individuals with failure times right-censored at τ_l; the individuals in S_l are known to have survived the interval $(\tau_{l-1}, \tau_l]$ but were subsequently lost to view. The overall likelihood function, for the general case with explanatory variables, is

$$L = \Pi_{l=1}^{m}[\Pi_{i \in R_l}\{\bar{F}(\tau_{l-1};\boldsymbol{x}_i) - \bar{F}(\tau_l;\boldsymbol{x}_i)\} \times \Pi_{i \in S_l}\bar{F}(\tau_l;\boldsymbol{x}_i)].$$

This is formally the same as Equation 4.2.5, and so it reduces to Equation 4.2.8 in terms of hazard contributions. In the absence of explanatory variables, L reduces to Equation 4.2.2, whence estimators for $\bar{F}$ and f follow as in Equation 4.2.3.

4.3 Continuous failure times: random samples

4.3.1 The Kaplan-Meier estimator

Consider a random sample (i.e., without explanatory variables) of continuous failure time observations generated from some unknown underlying survivor function $\bar{F}(t)$. A parametric model for $\bar{F}(t)$ is *not* assumed. The estimation procedure here is very similar to that outlined in Section 4.1. The difference is that observed failure times t_l here replace the fixed discrete times τ_l there.

Let the distinct observed failure times be $t_1 < t_2 < \ldots < t_m$. Suppose that r_l failures are observed at time t_l, and that s_l cases are right-censored during the interval $[t_l, t_{l+1})$ at times t_{ls} (known or unknown), where $t_{ls} \le t_{l,s+1}$ ($s = 1, \ldots, s_l$; $l = 0, \ldots, m$; $t_0 = 0$, $t_{m+1} = \infty$). The likelihood function analogous to Equation 4.2.1 is

$$L = \Pi_{l=1}^{m}\{\bar{F}(t_l-) - \bar{F}(t_l)\}^{r_l} \times \Pi_{l=0}^{m}\Pi_{s=1}^{s_l}\bar{F}(t_{ls}). \tag{4.3.1}$$

The maximum likelihood estimator $\hat{\bar{F}}$ of $\bar{F}$ must be discontinuous on the left at t_l, otherwise $\hat{\bar{F}}(t_l-) = \hat{\bar{F}}(t_l)$ which makes $L = 0$. Also, since $t_{ls} \ge t_l$, $\hat{\bar{F}}(t_{ls}) \le \hat{\bar{F}}(t_l)$ and so L is maximised with $\hat{\bar{F}}(t_{ls}) = \hat{\bar{F}}(t_l)$. The likelihood does not involve $\bar{F}(t)$ for $t_{ls_l} < t < t_{l+1}$ and we can take $\hat{\bar{F}}(t) = \hat{\bar{F}}(t_l)$ on this interval. Hence, $\hat{\bar{F}}$ is a discrete survivor function with steps at $t_1, \ldots, t_m$. Applying Equation 4.1.1, $\hat{\bar{F}}(t_l)$ can be expressed in the form $\Pi_{s=1}^{l}(1-\hat{h}_s)$, and $\hat{\bar{F}}(t_l-) - \hat{\bar{F}}(t_l)$ as $\hat{h}_l\Pi_{s=1}^{l-1}(1-\hat{h}_s)$, where the $\hat{h}_s$ are chosen to maximise

$$L = \Pi_{l=1}^{m}\{h_l\Pi_{s=1}^{l-1}(1-h_s)\}^{r_l} \times \Pi_{l=0}^{m}\{\Pi_{s=1}^{l}(1-h_s)\}^{s_l};$$

the terms $\Pi_{s=1}^{0}(.)$ are interpreted as 1 because the first is $\bar{F}(0) - \bar{F}(t_1) = h_1$ and the second is $\Pi_{s=1}^{0}\bar{F}(0) = 1$. Collecting terms as in Equation 4.2.2,

$$L = \Pi_{l=1}^{m}\{h_l^{r_l}(1-h_l)^{q_l-r_l}\}, \quad (4.3.2)$$

where

$$q_l = (r_l + s_l) + \ldots + (r_m + s_m) \text{ for } l = 1, \ldots, m;$$

taking $r_0 = 0$, q_0 is the overall sample size and q_l for $l > 0$ is the number of cases at risk at time t_l-. The likelihood function L attains its maximum when $h_l = r_l/q_l$. The **product limit estimator** of $\bar{F}(t)$ is then

$$\hat{\bar{F}}(t) = \Pi(1-\hat{h}_l) = \Pi(1-r_l/q_l), \quad (4.3.3)$$

where the product is taken over $\{l{:}t_l \le t\}$. The formula is also known as the **Kaplan-Meier estimator** after their much-referenced 1958 paper. It is a nonparametric (more correctly, distribution-free) estimator of the survivor function.

If $s_m > 0$, so that at least one failure time is known to exceed t_m, $r_m/q_m < 1$ and so $\hat{\bar{F}}(t_m) > 0$. The downward progress of $\hat{\bar{F}}(t)$ beyond t_m to its eventual demise at value 0 is usually taken to be undefined in this case. The censoring times t_{ls} have no effect on $\hat{\bar{F}}(t)$, which might seem a bit strange at first sight. However, as far as the lifetime distribution is concerned, nothing happens at time t_{ls} — it is only the observation process which is interrupted there. When there is no censoring, $s_l = 0$ for all l, so $\hat{h}_l = r_l/(r_l + \ldots + r_m)$ and $1 - \hat{h}_l = q_{l+1}/q_l$, and then

$$\hat{\bar{F}}(t) = q_{l+1}/q_l = (r_{l+1} + \ldots + r_m)/(r_1 + \ldots + r_m) \text{ for } t_l \le t < t_{l+1},$$

which is the usual **empirical survivor function**. If, in addition, the t_l are all distinct, then $r_l = 1$ for all l, $q_1 = m$, $q_{l+1} = m - l$, and $\hat{\bar{F}}(t_l) = 1 - l/m$.

An asymptotic variance calculation can be made using an informal argument based on Greenwood's formula Equation 4.2.4:

$$\text{var}\{\hat{\bar{F}}(t)\} \approx \hat{\bar{F}}(t)^2 \Sigma \hat{h}_l/\{q_l(1-\hat{h}_l)\} = \hat{\bar{F}}(t)^2 \Sigma r_l/\{q_l(q_l - r_l)\}. \quad (4.3.4)$$

More rigorous asymptotic theory has been given by Meier (1975), for the case of censorship at fixed, predetermined times, and by Breslow and Crowley (1974), for the case of random censoring.

4.3.2 *The integrated and cumulative hazard functions*

The integrated hazard function was defined in Equation 1.1.1 as $H(t) = -\log \bar{F}(t)$. This suggests the estimator

$$\hat{H}(t) = -\log \hat{\bar{F}}(t) = -\Sigma \log(1-\hat{h}_l) = -\Sigma \log(1 - r_l/q_l), \quad (4.3.5)$$

from Equation 4.3.3, with summation over $\{l:t_l \leq t\}$. From an informal argument based on Equation 4.3.4 and the delta method,

$$\text{var}\{\hat{H}(t)\} \approx \hat{\bar{F}}(t)^{-2}\text{var}\{\hat{\bar{F}}(t)\} \approx \Sigma\hat{h}_l/\{q_l(1-\hat{h}_l)\}.$$

If the $\hat{h}_l$ are small, the alternative, **Nelson-Aalen estimator**

$$\tilde{H}(t) = \Sigma\hat{h}_l = \Sigma r_l/q_l \tag{4.3.6}$$

is suggested by the approximation to $H(t)$ given in Section 4.1.1; $\tilde{H}(t)$ is, strictly speaking, the **cumulative hazard function**. In fact, the r_l/q_l will tend to be small early on, i.e., for smaller values of l, so $\hat{H}(t)$ and $\tilde{H}(t)$ will not differ much for the smaller values of t. The corresponding alternative estimator of the survivor function is $\tilde{\bar{F}}(t) = \exp\{-\tilde{H}(t)\}$. The variance of $\tilde{H}(t)$ can be approximated as

$$\text{var}\{\tilde{H}(t)\} \approx \Sigma\hat{h}_l(1-\hat{h}_l)/q_l;$$

this is similar to that of $\hat{H}(t)$ when the $\hat{h}_l$ are small.

Plots of $\hat{H}$ or $\tilde{H}$ are useful for both exploratory analysis and assessment of goodness of fit (Nelson, 1970).

4.4 *Continuous failure times: explanatory variables*

4.4.1 *Cox's proportional hazards model*

The observational setting here comprises observed failure times t_l ($l = 1, \ldots, m$), taken to be in increasing order, and individual cases of whom the ith is accompanied by a vector $\boldsymbol{x}_i$ of explanatory variables. The proportional hazards model (Cox, 1972) specifies the hazard function for the ith individual as

$$h(t;\boldsymbol{x}_i) = \psi_i h_0(t) \tag{4.4.1}$$

where $h_0(t)$ is some unspecified **baseline hazard function** and $\psi_i = \psi(\boldsymbol{x}_i;\beta)$; here, ψ is some positive function, often taken to be of the form $\psi(\boldsymbol{x};\beta) = \exp(\boldsymbol{x}^T\beta)$, incorporating the vector β of regression coefficients. Note that "proportional hazards" here has a different meaning from that defined for Competing Risks in Section 1.2.

The survivor function corresponding to $h(t;\boldsymbol{x})$ is given by

$$\bar{F}(t;\boldsymbol{x}) = \exp\left\{-\int_0^t h(s;\boldsymbol{x})ds\right\}.$$

Substitution of Equation 4.4.1 into this yields

$$\bar{F}(t;\mathbf{x}_i) = \bar{F}_0(t)^{\psi_i}, \tag{4.4.2}$$

where $\bar{F}_0(t)$ is the **baseline survivor function** corresponding to $h_0(t)$. The family of survivor functions in Equation 4.4.2 looks suspiciously like that of Lehmann alternatives (Section 1.4.4).

The model Equation 4.4.1 is **semi-parametric** in that the ψ_i are parametrically specified, but $h_0(t)$ is not. It will be seen in the next section that this specification, together with the partial likelihood, yields a major benefit: one can estimate and test β without having to do anything about the unknown $h_0(t)$. However, as always, such an assumption should not be taken in too carefree a manner. It requires that the only effect of the covariate is to multiply the hazard by a factor which is constant over time. For instance, it would not be strictly valid in the case of a transient treatment effect, where the hazard is reduced for a while but later resumes its untreated level, the reduction depending on the dosage x. Cox and Oakes (1984, Section 5.7) discussed such an example.

4.4.2 Cox's partial likelihood

Let $R(t)$ be the risk set at time $t-$ (Section 4.2.3). Then the probability that individual $a \in R(t)$ fails in the time interval $(t, t + dt]$ is $h(t;\mathbf{x}_a)dt$. We assume for the moment that the individuals have distinct failure times, i.e., there are no ties. Given the events up to time t_l-, and given that there is a failure at time t_l, the conditional probability that it is individual i_l, among those still at risk, who fails is

$$h(t_l;\mathbf{x}_{i_l})/\Sigma_l h(t_l;\mathbf{x}_a) = \psi_{i_l}/\Sigma_l \psi_a, \tag{4.4.3}$$

where the summation Σ_l is over $a \in R(t_l)$ and Equation 4.4.1 has been applied. Notice that $h_0(t)$ has been cancelled out: this nuisance parameter, or function, has been eliminated; it is deceased; this nuisance parameter is no more; it is an ex-nuisance parameter. By analogy with ordinary likelihood, take the product over $l = 1, \ldots, m$ of Equation 4.4.3: the result is the so-called **partial likelihood function**

$$P(\beta) = \Pi_{l=1}^{m}\{\psi_{i_l}/\Sigma_l \psi_a\}. \tag{4.4.4}$$

The censored failure times make their presence felt only in the denominators of the factors of $P(\beta)$, the individuals concerned being included in the appropriate risk sets. Cox (1972) argued that there is no further information in the data about β, the parameter of interest. This is because $h_0(t)$ could be very small between failure times and not small at the observed t_l. It is only at the times t_l that β can affect the outcome through the *relative* chances of failure between different individuals.

The function $P(\beta)$ provides estimates and tests for β analogous to those based on ordinary likelihoods. Thus, the estimator is found by maximising $P(\beta)$ or, equivalently, $\log P(\beta)$. Setting $d_l = \Sigma_l \psi_a$, we have

$$\log P(\beta) = \sum_{l=1}^{m} \{\log \psi_{i_l} - \log d_l\}$$

with first and second derivatives, respectively,

$$\begin{aligned} U(\beta) &= \partial \log P(\beta)/\partial(\beta) = \sum_{l=1}^{m} U_l(\beta), \\ V(\beta) &= -\partial^2 \log P(\beta)/\partial\beta^2 = \sum_{l=1}^{m} V_l(\beta) \end{aligned} \tag{4.4.5}$$

where the vector $U_l(\beta)$ has jth component $\partial \log\{\psi_{i_l}/d_l\}/\partial\beta_j$ and the matrix $V_l(\beta)$ has jkth element $-\partial^2 \log\{\psi_{i_l}/d_l\}/\partial\beta_j\partial\beta_k$. With the usual choice, $\psi_i = \exp(x_i^{\mathrm{T}}\beta)$, we have $U_l(\beta) = x_{i_l} - v_l$ and $V_l(\beta) = X_l - v_l v_l^{\mathrm{T}}$, where

$$v_l = d_l^{-1} \Sigma_l \psi_a x_a \quad \text{and} \quad X_l = d_l^{-1} \Sigma_l \psi_a x_a x_a^{\mathrm{T}}. \tag{4.4.6}$$

Under some regularity conditions, $V(\hat{\beta})^{-1}$ provides an estimate for the variance matrix of $\hat{\beta}$, and so follow hypothesis tests and confidence regions for β.

A computationally useful aspect of partial likelihood with $\psi_i = \exp(x_i^{\mathrm{T}}\beta)$ is that $\log P(\beta)$ is a convex function of β, and so has a unique maximum, $\hat{\beta}$. To see this, note that, for an arbitrary vector z,

$$-z^{\mathrm{T}} V(\beta) z = \sum_{l=1}^{m} z^{\mathrm{T}} V_l(\beta) z = \sum_{i=1}^{m} \{z^{\mathrm{T}} X_l z - (z^{\mathrm{T}} v_l)^2\} = \sum_{l=1}^{m} \{\Sigma_l \pi_a z_a^2 - (\Sigma_l \pi_a z_a)^2\},$$

where $z_a = z^{\mathrm{T}} x_a$ and $\pi_a = \psi_a/d_l$. Let $\bar{z}_a = \Sigma_l \pi_a z_a$ and note that $\Sigma_l \pi_a = 1$. Then

$$-z^{\mathrm{T}} V(\beta) z = \sum_{l=1}^{m} \Sigma_l \pi_a (z_a - \bar{z}_a)^2 \geq 0,$$

so $\log P(\beta)$ has a negative definite second derivative matrix as claimed. The numerical implications of this for maximising such functions are discussed in Appendix 1.

A useful extension to the analysis can be made via stratification. Suppose that there are groups of individuals in the data, arising from different subpopulations or strata, between which $h_0(t)$ is likely to differ. One can allow for this by constructing a separate partial likelihood function for each group and then taking the overall $P(\beta)$ as the product of these separate ones. The regression parameter β may be common to all groups or may differ between groups in any respect.

When there are tied values among the observed failure times, the expression for $P(\beta)$ is more complicated. Thus, if there are r_l observed failure times equal to t_l, the lth term in the product defining $P(\beta)$ should allow for all distinct subsets of r_l individuals from the risk set $R(t_l)$. This usually leads to an unwieldy computation, and an approximation due to Peto (1972) is to replace this lth term by $\Pi(\psi_{i_l}/d_l)$, where the product is taken over the r_l tied cases. Since d_l is the same for all these cases, the replacement term can be written as $(\Pi\psi_{i_l})/d_l^{r_l}$. This approximation will give reasonable results provided that the proportion of tied values in each $R(t_l)$ is small. Otherwise, it might be feasible to compute the partial likelihood contributions taking full account of all the combinatorial possibilities of r_l individuals from $R(t_l)$. Failing that, options like those listed in the following paragraph can be considered.

In the absence of explanatory variables, there is only the hazard function to estimate and this can be done parametrically, with a fully specified model, or nonparametrically, using the Kaplan-Meier estimator (Section 4.3.1). In order to deal with explanatory variables, the options are, broadly, (a) a fully parametric model for the hazards $h(t;\boldsymbol{x})$, or their equivalents, and (b) a model that separates the regression and hazard components, i.e., the $\boldsymbol{x}$ and the t parts, leaving the latter parametrically unspecified. In (b) it might be possible to find a combination of model and estimating function that allows the unspecified baseline hazards to be eliminated, as in Cox's combination of proportional hazards and partial likelihood. The full likelihood function needed for (a), and for (b) if the baseline hazards are to be estimated, is Equation 4.2.5 or Equation 4.2.8.

4.4.3 The baseline survivor function

The baseline survivor function $\bar{F}_0$ can be estimated non-parametrically by an extension of the product limit estimator (Section 4.3.1). Let t_l ($l = 1, \ldots, m$) be the distinct failure times in increasing order, with $t_0 = 0$ and $t_{m+1} = \infty$; let R_l be the set of individuals failing at time t_l, and let S_l likewise comprise the cases right-censored during $[t_l, t_{l+1})$. The likelihood function analogous to Equation 4.2.5 becomes, using Equation 4.4.2,

$$L = \Pi_{l=1}^{m}[\Pi_{i \in R_l}\{\bar{F}_0(t_l-)^{\psi_i} - \bar{F}_0(t_l)^{\psi_i}\} \times \Pi_{i \in S_l}\{\bar{F}_0(t_i)^{\psi_i}\}]. \qquad (4.4.7)$$

As for Equation 4.3.1, $\hat{\bar{F}}_0$ must jump at t_l- for each l, and $\hat{\bar{F}}_0(t) = \hat{\bar{F}}_0(t_l)$ for $t_l \le t < t_{l+1}$. Hence, $\hat{\bar{F}}_0(t_l) = \Pi_{s=1}^{l}(1 - \hat{h}_{0s})$ where the $\hat{h}_{0s}$ maximise

$$L = \Pi_{l=1}^{m}[\Pi_{i \in R_l}\{\Pi_{s=1}^{l-1}(1-h_{0s})^{\psi_i} - \Pi_{s=1}^{l}(1-h_{0s})^{\psi_i}\} \times \Pi_{i \in S_l}\Pi_{s=1}^{l}(1-h_{0s})^{\psi_i}]$$
$$= \Pi_{l=1}^{m}\{\Pi_{i \in R_l}(1-g_{si}) \times \Pi_{i \in R_l}\Pi_{s=1}^{l-1}g_{si} \times \Pi_{i \in S_l}\Pi_{s=1}^{l}g_{si}\};$$

we have written g_{si} here for $(1-h_{0s})^{\psi_i}$. By exactly the same reduction that led from Equation 4.2.6 to Equation 4.2.8, we obtain

$$L = \Pi_{l=1}^{m}[\Pi_{i \in R_l}\{1-(1-h_{0l})^{\psi_i}\} \times \Pi_{i \in Q_l}(1-h_{0l})^{\psi_i}], \qquad (4.4.8)$$

where $Q_l = R(t_l) - R_l$ as in Equation 4.2.7. Thus, the $\hat{h}_{0l}$ may be computed by maximisation of L, after substituting the partial likelihood estimate $\hat{\beta}$ for β. Note that L is of orthogonal form, being a product of m factors each involving just one h_{0l}. Thus, the maximisation problem can be reduced to a set of m one-dimensional searches (Appendix 1).

Alternatively, Equation 4.4.8 could be maximised jointly over β and the h_{0l} (Kalbfleisch and Prentice, 1980, Section 4.3). However, the resulting estimator for β might not be as well behaved computationally as the maximum partial likelihood estimator.

4.4.4 *Residuals*

Residuals can be computed via the probability integral transform, as described in Section 2.2.2. However, more sophisticated approaches have been developed in recent years. Barlow and Prentice (1988) and Therneau, Grambsch, and Fleming (1990) defined and employed so-called **martingale residuals**. These cover a wide class of residuals applicable to both parametric and semi-parametric Survival Analysis and also cater for time-dependent covariates. As a simple example suppose that H_i denotes the integrated hazard function of t_i. Then $H_i(t_i)$ has distribution exp(1), exponential with unit mean. This is just another way of saying that $\bar{F}(t_i;\theta)$ has distribution U(0, 1), noting that $\bar{F}(t_i;\theta) = \exp\{-H_i(t_i)\}$. Hence, the set $\{\hat{H}_i(t_i):i = 1, \ldots, n\}$, based on estimated H_i-functions, should resemble a random sample from exp(1). If t_i is right-censored, $\hat{H}_i(t_i)$ should be adjusted by adding 1, since this is the mean residual lifetime for an exp(1) variate. The form $\hat{H}_i(t_i) + 1 - c_i$, where c_i is a censoring indicator taking value 0 if t_i is censored and 1 otherwise, covers both censored and non-censored cases. More generally, consider $U_i(t) = \hat{H}_i\{\min(t_i, t)\} - c_i(t)$, where t represents time and $c_i(t) = I(t_i < t, c_i = 1)$; here $I(E)$ is the **indicator function** of the condition E, i.e., $I(E) = 1$ if E is satisfied and $I(E) = 0$ if not. This can be shown to be a continuous-time martingale (Section 8.10.3), and the papers cited show how to use such martingale residuals for investigations of goodness-of-fit, regression diagnostics, influence, and outlier detection.

4.4.5 Some theory for partial likelihood

Cox (1975) provided some more formal justification for the likelihood-like properties of this function, Oakes (1981) considered further aspects, and Wong (1985) continued with the task at a more rigorous level. However, the most elegant treatment is via martingale counting processes (Chapter 8): Andersen and Gill (1982) provided a general theory; Gill (1984) and Andersen and Borgan (1985) contain very readable accounts of this rather technical area. An introduction to this topic will be given in Chapter 8, but here we will just follow Cox and give an informal justification for inference based on the partial likelihood (4.4.4).

In general terms, a partial likelihood function is set up as follows. Suppose that $\theta = (\beta, \phi)$, where β is the parameter of interest and ϕ is a nuisance parameter; in the case of the proportional hazards model, ϕ is h_0, the baseline hazard function, of infinite dimension. Suppose further that the accumulating data sequence can be formulated as $D_l = (A_1, B_1, \ldots, A_l, B_l)$ for $l = 1, 2, \ldots$. Then the likelihood function can be written as

$$L(\theta) = f_\theta(D_m) = \Pi_{l=1}^{m} f_\theta(A_l, B_l | D_{l-1}) = \Pi_{l=1}^{m} f_\theta(A_l | D_{l-1}, B_l) f_\theta(B_l | D_{l-1})$$
$$= \Pi_{l=1}^{m} f_\theta(A_l | D_{l-1}, B_l) \times \Pi_{l=1}^{m} f_\theta(B_l | D_{l-1}) = P(\theta) Q(\theta),$$

say, where f_θ is a generic notation for a probability density or mass function. Now, this formulation will be useful if $P(\theta)$ is a function only of β, the parameter of interest, and if $Q(\theta)$ does not depend on β. In that case, we can ignore the "constant" factor $Q(\theta)$ in the likelihood factorisation $L(\theta) = P(\beta)Q(\phi)$ and just use $P(\beta)$ for inference about β. Normally, however, $Q(\theta)$ will depend to some extent on β and therefore contain some residual information about β. Cox's justification for ignoring $Q(\theta)$, in the case of the proportional hazards model, is that this residual information is unavailable because β and ϕ are inextricably entangled in $Q(\theta)$. This type of argument has a long history (e.g., Kalbfleisch and Sprott, 1970).

To see how this factorisation is applied to the proportional hazards setup, consider data of the type considered in Section 4.3.1. Thus, the distinct observed failure times are $t_1 < \ldots < t_m$ and the censoring times during $[t_l, t_{l+1})$ are t_{ls} $(s = 1, \ldots, s_l)$; also, assuming no ties, the individual who failed at time t_l has label i_l, and those censored during $[t_l, t_{l+1})$ have labels i_{ls} $(s = 1, \ldots, s_l)$. Let $C_l = \{t_{ls}, i_{ls}: s = 1, \ldots, s_l\}$ comprise the full record of censored cases during $[t_l, t_{l+1})$, i.e., the censoring times and the individuals involved. Now take $A_l = \{i_l\}$ and $B_l = \{C_{l-1}, t_l\}$ in the above factorisation of $L(\theta)$. Then, $f_\theta(A_l | D_{l-1}, B_l)$ is just the conditional probability in Equation 4.4.3, and so $P(\beta)$ here is the one in Equation 4.4.4.

The asymptotic properties of maximum partial likelihood estimators are generally similar to those of ordinary maximum likelihood estimators, as pointed out by Cox (1975). The partial score function is, as in Equation 4.4.5,

$$\boldsymbol{U}(\beta) = \partial \log P(\beta)/\partial\beta = \sum_{l=1}^{m} \partial \log f_\beta(A_l|D_{l-1}, B_l)/\partial\beta = \sum_{l=1}^{m} \boldsymbol{U}_l(\beta).$$

Under the "usual regularity conditions," we can differentiate the identity $\int f_\beta(A_l|D_{l-1}, B_l)dA_l = 1$ with respect to β under the integral sign (or summation sign in the case of a discrete A_l, as it is in the case of the proportional hazards partial likelihood) to obtain

$$\mathrm{E}_\beta\{\partial \log f_\beta(A_l|D_{l-1}, B_l)/\partial\beta | D_{l-1}, B_l\} = \mathrm{E}_\beta\{\boldsymbol{U}_l(\beta)|D_{l-1}, B_l\} = \mathbf{0},$$

where $\mathrm{E}_\beta(.)$ indicates that the expectation is taken with respect to the same β-value as in $f_\beta(.)$. (In formal terms, this shows that the $\boldsymbol{U}_l(\beta)$ form a sequence of martingale differences adapted to the filtration $\{D_{l-1}, B_l\}$, see Section 8.4.) Hence, $\mathrm{E}_\beta\{\boldsymbol{U}_l(\beta)\} = \mathbf{0}$ and so $\mathrm{E}_\beta\{\boldsymbol{U}(\beta)\} = \mathbf{0}$ for each β, i.e., $\boldsymbol{U}(\beta) = \mathbf{0}$ is an **unbiased estimating equation** for β.

Denote by β_0 the true value of β, and by $\mathrm{E}_0(.)$ expectation taken under β_0, and let $\mu(\beta) = \mathrm{E}_0\{\boldsymbol{U}(\beta)\}$. Suppose that, under β_0, (i) $\boldsymbol{U}(\beta)$ obeys a law of large numbers for each β, to the effect that it stabilizes around $\mu(\beta)$ as $m \to \infty$, (ii) $\mu(\beta)$ stays clear of $\mathbf{0}$ whenever $\beta \neq \beta_0$, and (iii) either $\boldsymbol{U}(\beta)$ or $\mu(\beta)$ obeys some suitable continuity condition. Then, from the general theory of estimating equations (Crowder, 1986), the maximum partial likelihood estimator $\tilde{\beta}$ will be **consistent** for β_0. In particular, for the basic proportional hazards model with $\psi_i = \exp(x_i^{\mathrm{T}}\beta)$, $\log P(\beta)$ is a differentiable convex function (Section 4.4.2), so conditions (ii) and (iii) will be within grasp.

The asymptotic distribution of $\tilde{\beta}$ can be investigated via the mean value theorem: we have, expanding about the true value β_0,

$$\mathbf{0} = \boldsymbol{U}(\tilde{\beta}) = \boldsymbol{U}(\beta_0) - \boldsymbol{V}(\beta_*)(\tilde{\beta} - \beta_0),$$

where $\boldsymbol{V}(\beta) = -\partial^2 P(\beta)/\partial\beta^2$ and β_* lies between β_0 and $\tilde{\beta}$. Rearranging, approximating $\boldsymbol{V}(\beta_*)$ by $\boldsymbol{V}(\beta_0)$, and assuming that $\boldsymbol{V}(\beta_0)$ stabilizes towards its expected value as $n \to \infty$, we have

$$\tilde{\beta} - \beta_0 \approx \boldsymbol{M}^{-1}\boldsymbol{U}(\beta_0),$$

where $\boldsymbol{M} = \mathrm{E}_0\{\boldsymbol{V}(\beta_0)\}$ must be non-singular; $\boldsymbol{M}$ could be described as the expected partial information matrix. The asymptotic distribution of $\tilde{\beta} - \beta_0$ is thus that of $\boldsymbol{M}^{-1}\boldsymbol{U}(\beta_0)$.

Since the realized value of $\boldsymbol{U}_l$ is determined by $\{D_{k-1}, B_k\}$ when $l < k$, we have

$$\mathrm{E}_\beta\{\boldsymbol{U}_l(\beta)\boldsymbol{U}_k(\beta) \mid D_{k-1}, B_k\} = \boldsymbol{U}_l(\beta) \times \mathrm{E}_\beta\{\boldsymbol{U}_k(\beta) \mid D_{k-1}, B_k\} = \boldsymbol{U}_l(\beta) \times \mathbf{0} = \mathbf{0},$$

so $E_\beta\{\boldsymbol{U}_l(\beta)\boldsymbol{U}_k(\beta)\} = \mathbf{0}$, i.e., the $\boldsymbol{U}_l$ are uncorrelated; this is a standard property of any martingale difference sequence. Again, by differentiating the identity $\int f_\beta(A_l|D_{l-1}, B_l)dA_l = 1$ a second time with respect to β under the integral sign, we obtain

$$\mathbf{0} = E_\beta\{\partial^2 \log f_\beta(A_l \mid D_{l-1}, B_l)/\partial\beta^2 + \boldsymbol{U}_l(\beta)\boldsymbol{U}_l(\beta)^T \mid D_{l-1}, B_l\},$$

i.e.,

$$E_\beta\{\boldsymbol{V}_l(\beta) \mid D_{l-1}, B_l\} = E_\beta\{\boldsymbol{U}_l(\beta)\boldsymbol{U}_l(\beta)^T \mid D_{l-1}, B_l\} = \text{var}_\beta\{\boldsymbol{U}_l(\beta) \mid D_{l-1}, B_l\}.$$

Hence,

$$\begin{aligned}\boldsymbol{M} &= E_0\{\boldsymbol{V}(\beta_0)\} = \Sigma E_0\{\boldsymbol{V}_l(\beta_0)\} = \Sigma E_0\{\boldsymbol{U}_l(\beta_0)\boldsymbol{U}_l(\beta_0)^T\} \\ &= E_0\{\boldsymbol{U}(\beta_0)\boldsymbol{U}(\beta_0)^T\} = \text{var}_0\{\boldsymbol{U}(\beta_0)\},\end{aligned}$$

where the fact that the $\boldsymbol{U}_l(\beta_0)$ are uncorrelated under β_0 has been used. To summarize, $\boldsymbol{U}(\beta_0)$ is the sum of m terms $\boldsymbol{U}_l(\beta_0)$, uncorrelated under β_0, and has covariance matrix $\boldsymbol{M}$. Under a mild condition to prevent the dominance of any one term in the sum, e.g., that the variances of the $\boldsymbol{U}_l(\beta_0)$ are bounded uniformly over l, $\boldsymbol{M}^{-1/2}\boldsymbol{U}(\beta_0)$ will tend as $m \to \infty$ to a normal distribution with mean $\mathbf{0}$ and unit covariance matrix $\mathbf{I}$. It then follows from above that $\tilde{\beta} \sim N(\beta_0, \boldsymbol{M}^{-1})$ and, in practice, $\boldsymbol{M}$ can be estimated as $\boldsymbol{V}(\tilde{\beta})$.

4.5 Discrete failure times again

Suppose, as in Section 4.1.1, that the defined failure times are $0 = \tau_0 < \tau_1 < \ldots < \tau_m$. Denote by R_l the set of r_l individuals failing at time τ_l and by S_l the set of s_l individuals right-censored at τ_l. With the ith case is associated the vector $\boldsymbol{x}_i$ of explanatory variables.

4.5.1 Proportional hazards

Assume a proportional hazards model expressed as

$$\bar{F}(t, \boldsymbol{x}_i) = \bar{F}_0(t)^{\psi_i},$$

where $\psi_i = \psi(\boldsymbol{x}_i;\beta)$. Writing $h_s(\boldsymbol{x}_i)$ for the discrete hazard contribution $h(\tau_s;\boldsymbol{x}_i)$, and h_{0s} for $h_0(\tau_s)$, we have

$$\bar{F}_0(\tau_l) = \Pi_{s=1}^{l}(1 - h_{0s}), \quad \bar{F}(\tau_l, \boldsymbol{x}_i) = \Pi_{s=1}^{l}\{1 - h_s(\boldsymbol{x}_i)\},$$

which entails

$$h_s(x_i) = 1 - (1 - h_{0s})^{\psi_i}. \tag{4.5.1}$$

Unfortunately, Equation 4.5.1 does not yield cancellation of the h_{0s} when a partial likelihood is set up along the lines of Equation 4.4.4, so we cannot follow that route to estimate β free of the baseline hazards. The alternative route, setting up the proportional hazards model directly as $h_s(x_i) = \psi_i h_{0s}$, would lead to elimination of the h_{0s} in Equation 4.4.3 as before. However, again unfortunately, Equation 4.4.3 is not the correct expression of the intended conditional probability, as will be explained below in Section 4.5.2.

The likelihood function Equation 4.2.8 is, for the present case,

$$L = \Pi_{l=1}^{m}[\Pi_{i \in R_l}\{1 - (1 - h_{0l})^{\psi_i}\} \times \Pi_{i \in Q_l}(1 - h_{0l})^{\psi_i}]. \tag{4.5.2}$$

This likelihood, essentially the same as Equation 4.4.8, can now be maximized jointly over β and the h_{0l}. For standard asymptotic theory to apply there needs to be a finite, preferably small, number of parameters. For unrestricted h_{0l} this would entail finite m; otherwise, the h_{0l} would have to be restricted by being expressed as functions of a finite set of parameters.

4.5.2 *Proportional odds*

An alternative model, which can be used to advantage with the partial likelihood approach, was proposed by Cox (1972). We begin with a more careful derivation of the conditional probability Equation 4.4.3; this is the probability that individual i_l fails at time τ_l and the others in the risk set $R(\tau_l)$ do not, given that there is one failure at τ_l. Writing $h_l(x_i)$ for $h(\tau_l;x_i)$, this is

$$\begin{aligned} &h_l(x_{i_l})\Pi_{a \neq i_l}\{1 - h_l(x_a)\} \div \Pi_{i \in R(t_l)}[h_l(x_i)\Pi_{a \neq i}\{1 - h_l(x_a)\}] \\ &= [h_l(x_{i_l})/\{1 - h_l(x_{i_l})\}] \div \Pi_{i \in R(t_l)}[h_l(x_i)/\{1 - h_l(x_i)\}]. \end{aligned} \tag{4.5.3}$$

In the continuous case $h_l(x_i)$ is replaced by $h(t_l;x_i)dt$, and then, as $dt \to 0$, Equation 4.4.3 is recovered.

Cancellation of the baseline hazards in Equation 4.5.3 will be achieved by assuming that

$$h_l(x)/\{1 - h_l(x)\} = \psi(x;\beta)h_{0l}/(1 - h_{0l}), \tag{4.5.4}$$

where $h_{0l} = h_0(\tau_l)$ is the baseline hazard contribution $\text{pr}(T = \tau_l \mid T \geq \tau_l)$. The ratio $h_{0l}/(1 - h_{0l})$ is the conditional odds on the event $\{T = \tau_l\}$ under the baseline hazard. Thus, Equation 4.5.4 represents proportional odds (Section 1.3.3). With the usual choice, $\psi(x;\beta) = \exp(x^T\beta)$, this boils down to a logit model:

$$\log[h_l(x)/\{1 - h_l(x)\}] = \alpha_l + x^T\beta,$$

where $\alpha_l = \log\{h_{0l}/(1 - h_{0l})\}$. Under Equation 4.5.4 the partial likelihood takes the same form as in Equation 4.4.4. In fact, the only difference here is to replace the proportional hazards model Equation 4.4.1 by the proportional odds model Equation 4.5.4.

Tied failure times are very likely to occur in the discrete case. Formally, if individuals $i_1, \ldots, i_r$ all fail at time τ_l, the lth contribution to the partial likelihood is $\psi_{i_1}\cdots\psi_{i_r}/\Sigma_l\psi_{a_1}\cdots\psi_{a_r}$, where the summation is over all subsets $(a_1, \ldots, a_r)$ of individuals from $R(\tau_l)$; this form arises as before, through Equation 4.5.3 and Equation 4.5.4. It might be feasible to compute $P(\beta)$ fully in this way (Gail et al., 1981). Otherwise, an approximate adjustment for (not too many) ties can be made as described in Section 4.4.2 for the continuous case. If there is a substantial proportion of ties in the data, and exact computation is not feasible, one must settle for the kinds of alternative options described in Section 4.4.2. For this, the appropriate likelihood function is Equation 4.2.5 or Equation 4.2.8.

The baseline hazards h_{0l} in Equation 4.5.4 can be estimated as follows. Having obtained the estimates $\hat{\psi}_i = \psi(x_i;\hat{\beta})$ from the partial likelihood, we can use Equation 4.5.4 to substitute $h_l(x_i)$ for $h_l(x_i)$ in Equation 4.2.8. This gives

$$L = \Pi_{l=1}^{m}[\Pi_{i \in R_l}\{\hat{\psi}_i h_{0l}/(1 - h_{0l})\} \times \Pi_{i \in R(t_l)}\{\hat{\psi}_i h_{0l}/(1 - h_{0l} + \hat{\psi}_i h_{0l})\}],$$

which can be maximised over the h_{0l}. As in Equation 4.4.8, L is of orthogonal form in the h_{0l}, thus simplifying the computation considerably.

4.6 Time-dependent covariates

The procedures described above can be extended to cope with time-dependent covariates. Let the value of the covariate vector be $x_i(t)$ for the ith individual at time t, denote by $x_i(0, t)$ the covariate history over the time period $(0, t)$, and let $x_i = x_i(0, \infty)$. In the previous development $x_i(t)$ was constant over time. More generally, some components of $x_i(t)$ might be constant and others not. A distinction is made between **external covariates** and **internal covariates** (Kalbfleisch and Prentice, 1978, Section 5.3).

An external covariate $x_i(t)$ is determined independently of the progress of the individual under observation. Thus, $x_i(t)$ can represent a pre-specified regime of conditions to which the ith individual is subjected, some function of clock time, or a non-predetermined stochastic process whose probability distribution does not involve the parameters of the failure time model; environmental conditions might be a case in point. In such cases it is natural to condition the inferences on the observed realization of $x_i(t)$.

An internal covariate $x_i(t)$ is generated as part of the individual's development over time. An example is where $x_i(t)$ gives information on the state of the individual or system at time t and therefore can have a very close bearing on the probable failure time. This case is often more difficult to interpret because, for instance, the value of the internal covariate might essentially determine whether failure has occurred or not and hence effectively remove some or all of the stochastic structure of T. On the other hand, $x_i(t)$, being much more relevant to imminent failure than $x_i(0)$, might be an essential ingredient of a useful model.

When there are time-dependent covariates the continuous-time hazard function becomes

$$h(t;x_i) = h\{t;x_i(0, t)\} = \lim_{\delta\to 0}\text{pr}\{T \le t + \delta \,|\, T > t, x_i(0, t)\} \qquad (4.6.1)$$

and the corresponding survivor function is given formally as

$$\bar{F}(t;x_i) = \exp\left[-\int_0^t h\{s;x_i(0, s)\}ds\right]. \qquad (4.6.2)$$

A natural extension of the proportional hazards model Equation 4.4.1 for time-dependent covariates is

$$h(t;x_i) = h_0(t)\psi\{x_i(t);\beta\}. \qquad (4.6.3)$$

This formally depends on the process x_i only through its value $x_i(t)$ at time t, so one has to specify which functions of x_i are to be represented, e.g., an integral over $(0, t)$ for some cumulative effect. The relation 4.4.2, expressing $F(t;x_i)$ in terms of a baseline survivor function corresponding to h_0, no longer holds when ψ varies with t; this is simply because ψ_i cannot be taken outside the integral in Equation 4.6.2. Also, the hazards in Equation 4.6.3 might no longer be proportional: the hazard ratio between cases i and j is

$$h(t;x_i)/h(t;x_j) = \psi\{x_i(t);\beta\}/\psi\{x_j(t);\beta\},$$

and this is likely to vary with t. This can be put to use as a general approach to detecting departures from proportionality: the possible departure in question is built into the model, and then the corresponding coefficients are parametrically tested.

Example 4.4

Take $h(t;x_i) = h_0(t)\psi\{x_i(t);\beta\}$ with $h_0(t) = \alpha t^{\gamma-1}$ (Weibull form), $\psi\{x_i(t);\beta\} = \exp\{\beta x_i(t)\}$, and $x_i(t) = (x_i\log t)$. Then,

$$h(t;x_i) = \alpha t^{\gamma-1}\exp(\beta x_i\log t) = \alpha t^{\gamma-1+\beta x_i},$$

and

$$\bar{F}(t;x_i) = \exp\{-\alpha_i t^{\gamma + \beta x_i}\},$$

where $\alpha_i = \alpha/(\gamma + \beta x_i)$. The baseline hazard obtains when $x_i = 0$, and the corresponding survivor function is $\bar{F}_0(t) = \exp(-\alpha t^\gamma/\gamma)$. However, unlike in Equation 4.4.2, $\bar{F}(t;x_i) \neq \bar{F}_0(t)^{\psi_i}$ here. Also, $h(t;x_i)/h(t;x_j) = t^{\beta(x_i - x_j)}$ varies with t, so the hazards are not proportional. ■

A time-dependent covariate can also be usefully employed to deal with cases where there is some intervention, like treatment, repair, or maintenance, which changes the hazard function for an individual.

Example 4.5

Suppose that the hazard for individual i is $h_0(t)\psi_1(x_i)$ before intervention, and $h_0(t)\psi_2(x_i)$ afterwards. Then, with intervention taking place at time τ_i, we may take the full hazard function to be $h_0(t)\psi(x_i;t)$ with

$$\psi(x_i;t) = I(t < \tau_i)\psi_1(x_i) + I(t \geq \tau_i)\psi_2(x_i),$$

$I(.)$ representing the indicator function. An assessment of the effect of intervention can now be made by comparing the estimates of $\psi_1(x)$ and $\psi_2(x)$ over the sample. ■

Example 4.6

Suppose that, instead of the sudden change from ψ_1 to ψ_2 at time τ_i in the previous example, the transition is smoother. Suppose that the covariate function is ψ_1 initially and then, as time goes on, moves over to ψ_2 in a continuous manner. This might reflect a cumulative effect of the covariates, or steadily changing conditions. A suitable model would be

$$\psi(x_i;t) = w(t)\psi_1(x_i) + \{1 - w(t)\}\psi_2(x_i),$$

where w is a function continuously decreasing from 1 at $t = 0$ to 0 at $t = \infty$, e.g., $w(t) = e^{-\gamma t}$. More generally, w might also depend on x_i. ■

In the case of time-dependent covariates the partial likelihood Equation 4.4.4 becomes

$$P(\beta) = \Pi_{l=1}^{m}\{\psi_{i_l l}/\Sigma_l \psi_{al}\},$$

where ψ_{al} is written for $\psi\{x_a(t_l);\beta\}$. Notice that, in the lth term of this product, the covariate vectors for all individuals involved are evaluated at time t_l. This extended form of $P(\beta)$ can be employed in the same way as described in Section 4.4.2 to obtain an estimate for β and associated asymptotic inference. However, a baseline survivor function cannot be estimated in the manner of Section 4.4.3 because, as noted above, the fundamental relation 4.4.2 fails here.

chapter 5

Discrete failure times in Competing Risks

It might be thought a little strange that this chapter on discrete failure times, surely a rather specialised topic, should precede the one on continuous failure times. The reason is that the non-parametric methods for the continuous case rely on the machinations and manipulations for the discrete case, covered here in Section 5.5. It is exactly the same situation as that which called for the material of Section 4.1 to come first in that chapter.

5.1 Basic probability functions

In this first section we will just set out the basic probability functions. We consider the general case where the set of possible failure times is $0 = \tau_0 < \tau_1 < \ldots < \tau_m$; m and τ_m may each be finite or infinite. In many situations it is sufficient to take $\tau_l = l$ ($l = 0, 1, \ldots$), i.e., integer-valued failure times. However, when we come on to likelihood functions in Section 5.5, the generality is necessary in order to smooth the transition to the continuous-time case dealt with in Chapter 6.

The sub-distribution and sub-survivor functions are defined as in the continuous-time case (Section 1.1):

$$F(c, t) = \text{pr}(C = c, T \leq t), \quad \bar{F}(c, t) = \text{pr}(C = c, T > t);$$

C is the failure cause and T is the discrete failure time. Thus,

$$F(c, t) + \bar{F}(c, t) = \text{pr}(C = c) = p_c.$$

Zero lifetimes will be discounted by taking $\bar{F}(c, 0) = p_c$ for each c. The overall system distribution and survivor functions are given by

$$F(t) = \text{pr}(T \le t) = \sum_c F(c, t), \; \bar{F}(t) = \text{pr}(T > t) = \sum_c \bar{F}(c, t).$$

The discrete sub-density and overall system density functions are defined for $l = 1, \ldots, m$ by

$$f(c, \tau_l) = \text{pr}(C = c, T = \tau_l) = \bar{F}(c, \tau_{l-1}) - \bar{F}(c, \tau_l),$$
$$f(\tau_l) = \text{pr}(T = \tau_l) = \bar{F}(\tau_{l-1}) - \bar{F}(\tau_l) = \sum_c f(c, \tau_l);$$

if t is not equal to one of the τ_l, then $f(c, t) = 0$. Also, $f(c, 0) = 0$ and, for $l > 1$,

$$\bar{F}(c, \tau_l) = \sum_{s=l+1}^{m} f(c, \tau_s) = p_c - \sum_{s=1}^{l} f(c, \tau_s).$$

The sub-hazard and overall hazard functions are defined for $l = 1, \ldots, m$ as

$$h(c, \tau_l) = f(c, \tau_l)/\bar{F}(\tau_{l-1}),$$
$$h(\tau_l) = f(\tau_l)/\bar{F}(\tau_{l-1}) = \sum_c h(c, \tau_l);$$

for $l = 0$, $h(c, \tau_0) = 0$. At the last possible failure time, τ_m, we have $\bar{F}(\tau_m) = 0$, $\bar{F}(\tau_{m-1}) = f(\tau_m)$, and $h(\tau_m) = 1$. The following relations hold for $l = 1, \ldots, m$:

$$\bar{F}(\tau_l) = \Pi_{s=0}^{l}\{1 - h(\tau_s)\}, \; f(\tau_l) = h(\tau_l)\Pi_{s=0}^{l-1}\{1 - h(\tau_s)\}.$$

"Proportional hazards" obtains when $h(c, \tau_l)/h(\tau_l)$ is independent of l for each c, and Theorem 1.1 goes through as before: thus, proportional hazards is equivalent to independence of T and C, the time and cause of failure.

Example 5.1 Discrete version of Example 1.2

Consider the sub-survivor function $\bar{F}(j, t) = \pi_j \rho_j^t$ ($j = 1, \ldots, p$; $t = 1, \ldots, m = \infty$). This represents a mixture of geometric distributions: the set-up can be interpreted as first choosing a component, the jth with probability π_j, and then observing that component's geometric failure time. The discrete sub-densities are $f(j, t) = \pi_j \rho_j^{t-1}(1 - \rho_j)$, the system survivor function is $\bar{F}(t) = \sum_{j=1}^{p} \pi_j \rho_j^t$ with discrete density $f(t) = \sum_{j=1}^{p} \pi_j \rho_j^{t-1}(1 - \rho_j)$, and the marginal probabilities p_j for C are the π_j. The conditional distributions are defined by $\text{pr}(T > t \mid \text{cause } j) = \rho_j^t$ (geometric) and

$$\text{pr}(\text{cause } j \mid T = t) = \pi_j \rho_j^{t-1} / \sum_l \pi_l \rho_l^{t-1}(1 - \rho_l).$$

Thus, C and T are independent if and only if the ρ_j are all equal. The sub-hazards are

$$h(j, t) = \pi_j \rho_j^{t-1}(1 - \rho_j) / \sum_l \pi_l \rho_l^{t-1}$$

and the system hazard is

$$h(t) = \sum_j \pi_j \rho_j^{t-1}(1 - \rho_j) / \sum_l \pi_l \rho_l^{t-1}.$$

The condition for proportional hazards is that the ρ_j are all equal, the same as for independence of C and T. In that case, when $\rho_j = \rho$ for all j, $h(j, t) = \pi_j(1 - \rho)$ and $h(t) = 1 - \rho$ are independent of t. ■

5.2 *Latent failure times*

The theory will be developed now from the basis of an underlying joint survivor function $\bar{G}(t)$ of a vector $\boldsymbol{T} = (T_1, \ldots, T_p)$ of discrete latent failure times. Each T_j has as its range of possible values $0 = \tau_0 < \tau_1 < \ldots < \tau_m$, and $T = \min\{T_1, \ldots, T_p\}$. The marginal hazard function of T_j is defined as $h_j(t) = g_j(t)/\bar{G}_j(t-)$, $\bar{G}(t)$ being the marginal survivor function of T_j and $g_j(t)$ its marginal density. In the discrete-time case, we must, realistically, allow for the possibility of ties among the T_j. Thus, we extend the definition of C from a single cause to a multiple cause, or **configuration** of causes. For example, if components j, k, and l fail simultaneously, and before any others, we define $T = T_j = T_k = T_l$ and $C = \{j, k, l\}$. Hence, C can now take any one of $2^p - 1$ possible values, these being the non-empty subsets of $\{1, \ldots, p\}$. The functions $F(c, t)$ and $\bar{F}(c, t)$ are thus augmented, with c representing any failure configuration, not necessarily just a single index.

A modified version of Tsiatis' (1975) Theorem 3.1 can be established for discrete failure times. We will need the following difference operator. Let $w(\boldsymbol{t})$ be a scalar function of $\boldsymbol{t} = (t_1, \ldots, t_p)$ and define the operator ∇_1 by

$$\nabla_1(j)w(\boldsymbol{t}) = w(t_1, \ldots, t_j, \ldots, t_p) - w(t_1, \ldots, t_j-, \ldots, t_p);$$

thus, $\nabla_1(j)$ performs backward differencing on the jth component of $\boldsymbol{t}$. The general s-dimensional version can now be defined for $s \le p$ by

$$\nabla_s(\boldsymbol{c})w(\boldsymbol{t}) = \Pi_{j=1}^{s} \nabla_1(c_j)w(\boldsymbol{t}),$$

where the components of the vector $\boldsymbol{c} = (c_1, \ldots, c_s)$ are distinct elements of the set $\{1, \ldots, p\}$. Thus, ∇_s differences $w(\boldsymbol{t})$ on components $c_1, \ldots, c_s$, the order being immaterial. For example, if $p = 4$ and $\boldsymbol{c} = (1, 3)$, then $s = 2$ and

$$\nabla_s(\boldsymbol{c})w(\boldsymbol{t}) = w(t_1-, t_2, t_3-, t_4) - w(t_1-, t_2, t_3, t_4)$$
$$- w(t_1, t_2, t_3-, t_4) + w(t_1, t_2, t_3, t_4).$$

We have $\bar{F}(t) = \bar{G}(t\mathbf{1}_p)$ and the modified form of Theorem 3.1 states that

$$f(c, t) = (-1)^s \nabla_s(\boldsymbol{c})\bar{G}(t\mathbf{1}_p), \tag{5.2.1}$$

where s is the number of components in configuration c and $\boldsymbol{c}$ is the vector of their indices. Note that if t is not one of the τ_l then Equation 5.2.1 gives $f(c, t) = 0$.

Peterson's (1976) bounds (Theorem 3.2) hold as before, but with $\sum_{j=1}^{p}$ now replaced by $\sum_c$, summation over all configurations.

The notation $\bar{h}(c, t) = f(c, t)/\bar{F}(t)$ will be found useful: the $\bar{h}(c, t)$ are not themselves hazard functions, but are related to the proper ones $h(c, t)$ via

$$\bar{h}(c, t) = h(c, t)\bar{F}(t-)/\bar{F}(t) = h(c, t)/\{1 - h(t)\}.$$

Another way of looking at the relationship between these functions is to note that

$$h(c, \tau_l) = f(c, \tau_l)/\sum_{s=1}^{m} f(\tau_s), \quad \bar{h}(c, \tau_l) = f(c, \tau_l)/\sum_{s=l+1}^{m} f(\tau_s),$$

i.e., $\bar{h}(c, \tau_l)$ is like $h(c, \tau_l)$ but with the term $f(\tau_l)$ missing in the denominator. Unlike $h(c, \tau_l)$, $\bar{h}(c, \tau_l)$ can exceed unity; unlike $h(c, \tau_l)$ it is not a conditional probability. We will also make use of the analogous functions $\bar{h}_j(t) = g_j(t)/\bar{G}_j(t)$ and

$$\bar{h}(t) = f(t)/\bar{F}(t) = \sum_c \bar{h}(c, t) = h(t)/\{1 - h(t)\} = \text{pr}(T = t \mid T \geq t)/\text{pr}(T \neq t \mid T \geq t).$$

The last expression shows that $\bar{h}(t)$ is the conditional odds for failure at time t, given survival to time t. Consequently, the $\bar{h}(c, t)$ may be interpreted as **sub-conditional-odds ratios**. At the upper end point τ_m of the T-distribution $h(t)$ and $\bar{h}(t)$ respectively take the values 1 and ∞, both for the first time.

The case of independent risks is defined by $\bar{G}(\boldsymbol{t}) = \Pi_{j=1}^{p}\bar{G}_j(t)$, as for continuous failure times. The convenient notation $f(j, t)$ will be used for $f(\{j\}, t)$ whenever the configuration is a simple index; likewise, $\bar{F}(j, t)$, $h(j, t)$, and p_j, respectively, stand for $\bar{F}(\{j\}, t)$, $h(\{j\}, t)$, and $p\{j\}$. The following theorem is the discrete-time analogue of Theorem 3.3.

Theorem 5.1

The following implications hold for the statements given below: (i) ⇒ (ii) and (ii) ⇒ the rest.

(i) independent risks obtains;

(ii) $\bar{h}(c, t) = \Pi_{j \in c}\bar{h}_j(t)$ for all (c, t), in particular, $\bar{h}(j, t) = \bar{h}_j(t)$ for $j = 1, \ldots, p$;

(iii) $\bar{h}(c, t) = \Pi_{j \in c}\bar{h}(j, t)$ for all (c, t);

(iv) $\bar{G}_j(t) = \Pi_{s=1}^{l(t)}\{1 + \bar{h}(j, \tau_s)\}^{-1}$, where $l(t) = \max\{l : \tau_l \le t\}$;

(v) $\bar{F}(t) = \Pi_{s=1}^{l(t)}\Pi_{j=1}^{p}\{1 + \bar{h}_j(\tau_s)\}^{-1}$;

(vi) $\bar{G}(t\mathbf{1}_p) = \Pi_{j=1}^{p}\bar{G}_j(t)$, i.e., $\bar{R}(t\mathbf{1}_p) = 1$ (for $\bar{R}$ see Section 3.1.1);

(vii) $f(c, t) = \{\Pi_{j \in c}\, g_j(t)\}\{\Pi_{j \notin c}\bar{G}_j(t)\}$.

Proof

Under (i), ties occur only through chance coincidence of the marginal processes, not through any additional interaction between them. Thus,

$$\begin{aligned} f(c, t) &= \text{pr}[\{\cap_{j \in c}(T_j = t)\}\cap\{\cap_{j \notin c}(T_j > t)\}] \\ &= \{\Pi_{j \in c}g_j(t)\} \times \{\Pi_{j \notin c}\bar{G}_j(t)\} \\ &= [\Pi_{j \in c}\{g_j(t)/\bar{G}_j(t)\}] \times [\{\Pi_{j=1}^{p}\bar{G}_j(t)\}] = \{\Pi_{j \in c}\bar{h}_j(t)\}\bar{F}(t), \end{aligned}$$

and (ii) follows. Statement (iii) follows immediately from (ii). For (iv) substitute

$$\bar{h}_j(t) = \{\bar{G}_j(t-)/\bar{G}_j(t)\} - 1$$

into the particular case of (ii). This yields the recurrence relation

$$\bar{G}_j(t) = \bar{G}_j(t-)\{1 + \bar{h}(j, t)\}^{-1}$$

from which (iv) follows, after setting $t = \tau_l$ and noting that $\bar{G}_j(\tau_l-) = \bar{G}_j(\tau_{l-1})$. For (v), sum (ii) over c: the left hand side gives

$$\sum_c \bar{h}(c, t) = \bar{h}(t) = \{\bar{F}(t-)/\bar{F}(t)\} - 1,$$

and the right hand side gives

$$\sum_c \prod_{j \in c} \bar{h}_j(t) = \Pi_{j=1}^{p}\{1 + \bar{h}_j(t)\} - 1.$$

Hence,

$$\bar{F}(t) = \bar{F}(t-) \div \Pi_{j=1}^{p}\{1 + \bar{h}_j(t)\}$$

and the result follows as for (iv). For (vi) note that (ii), (iv), and (v) together imply that $\bar{F}(t) = \Pi_{j=1}^{p}\bar{G}_j(t)$ and that $\bar{F}(t) = \bar{G}(t\mathbf{1}_p)$. For (vii) we have, using (ii),

$$\begin{aligned} f(c, t) &= \bar{F}(t)\bar{h}(c, t) = \bar{F}(t)\Pi_{j \in c}\{g_j(t)/\bar{G}_j(t)\} \\ &= \bar{R}(t\mathbf{1}_p)\{\Pi_{j \in c}g_j(t)\}\{\Pi_{j \notin c}\bar{G}_j(t)\}, \end{aligned}$$

and the result now follows from (vi). ■

In part (ii) of the theorem, the equality $\bar{h}(j, t) = \bar{h}_j(t)$ looks suspiciously like the Makeham assumption (Section 3.4.1). Proportional hazards cannot hold under (ii) because setting $h(c, t) = p_c h(t)$ in (ii) gives

$$p_c/\Pi_{j \in c}p_j = \bar{h}(t)^{|c|-1},$$

where $|c|$ is the number of single indices in configuration c; unless $|c| = 1$, one side varies with t and the other does not.

Part (iv) of the theorem says that the set of $\bar{h}(j, t)$, or equivalently the set $\{\bar{F}(j, t) : j = 1, \ldots, p\}$ of sub-survivor functions, determines the set of marginals $\bar{G}_j(t)$, i.e., the crude risks determine the net ones. Part (vi) says that "independence," i.e., $\bar{G}(\boldsymbol{t})$ equal to $\Pi_j\bar{G}_j(t_j)$, holds along the diagonal line $\boldsymbol{t} = t\mathbf{1}_p$. Part (vii) says that $f(c, t)$ is deceptively expressible in the same form that it would take if the risks acted independently. A similar point was made after the corresponding Theorem 3.3, which deals with proper hazard functions and simple configurations.

The theorem shows that (ii) is satisfied by any independent-risks model. That it is satisfied by some dependent-risks models but not by others is shown by the examples below in Section 5.3. More generally, such systems can be constructed in principle by starting with an independent-risks model and then shifting probability mass around in the style of Theorem 4 of Crowder (1991). Thus, one can change the joint $\bar{G}(\boldsymbol{t})$ without disturbing the sub-survivor functions $\bar{G}(j, t)$ nor the marginals $\bar{G}_j(t)$. As a result, the ratio $\bar{h}(c, t)/\Pi_{j \in c}\bar{h}_j(t)$ retains its original value, 1, whereas $\bar{R}(\boldsymbol{t})$ is moved away from 1 at one or more points off the diagonal $\boldsymbol{t} = t\mathbf{1}_p$. A similar construction can be used to show that (iii) does not imply (ii): probability mass can be shifted so that the $h(c, t)$ are unaltered, so (iii) is retained, but any particular marginal is altered, so (ii) is not retained.

Lemma 5.2

Under the condition $\bar{h}(j, t) = \bar{h}_j(t)$ $(j = 1, \ldots, p)$, proportional hazards obtains for single-index configurations, i.e., $h(j, t) = p_j h(t)$, if and only if

$$\bar{G}_j(t) = \Pi_{s=1}^{l(t)}\{1 + p_j h(\tau_s)\}^{-1}, \tag{5.2.2}$$

where $l(t) = \max\{l : \tau_l \leq t\}$.

Proof

The limited proportional hazards condition is equivalent to $\bar{h}(j, t) = p_j \bar{h}(t)$. Substituting this into (iv) of Theorem 5.1 gives Equation 5.2.2. Conversely, from Equation 5.2.2

$$g_j(t) = \bar{G}_j(t-) - \bar{G}_j(t) = \bar{G}_j(t)[\{1 + p_j \bar{h}(t)\} - 1],$$

from which $\bar{h}_j(t) = p_j \bar{h}(t)$ follows immediately. ■

In the case covered by the lemma the $\bar{G}_j(t)$, and hence $\bar{G}(t)$, are expressed in terms of the single function $\bar{h}(t)$, or equivalently $\bar{F}(t)$, alone. This is the discrete counterpart of Lemma 3.4.

5.3 *Some examples based on Bernoulli trials*

In this section some discrete counterparts of certain joint continuous failure time distributions will be examined. The basic construction is via replacement of the exponential distribution by the geometric, as illustrated in Example 5.1.

Example 5.2 Discrete version of Freund (1961)

Two sequences of Bernoulli trials are in progress, side by side in synchrony. After a failure in one sequence, the other continues with success probability changed from π_j to ρ_j ($j = 1, 2$).

The marginal densities associated with this model are $g_j(t) = \pi_j^{t-1}(1 - \pi_j)$ ($j = 1, 2$; $t = 1, \ldots, m = \infty$), and the sub-densities are

$$\begin{aligned} f(j, t) &= \pi_j^{t-1}(1 - \pi_j)\pi_k^t \qquad (k = 3 - j), \\ f(\{1, 2\}, t) &= \pi_1^{t-1}(1 - \pi_1)\pi_2^{t-1}(1 - \pi_2). \end{aligned} \tag{5.3.1}$$

The joint density of $\boldsymbol{T} = (T_1, T_2)$ is

$$g(\boldsymbol{t}) = \begin{cases} \pi_1^{t_1 - 1}(1 - \pi_1)\pi_2^{t_1}\rho_2^{t_2 - t_1 - 1}(1 - \rho_2) & \text{for } t_1 < t_2, \\ \pi_2^{t_2 - 1}(1 - \pi_2)\pi_1^{t_2}\rho_1^{t_1 - t_2 - 1}(1 - \rho_1) & \text{for } t_1 > t_2, \\ \pi_1^{t_1 - 1}(1 - \pi_1)\pi_2^{t_2 - 1}(1 - \pi_2) & \text{for } t_1 = t_2. \end{cases}$$

Note that $g(\boldsymbol{t}) \neq g_1(t_1)g_2(t_2)$ generally, so T_1 and T_2 are not independent (unless $\rho_1 = \pi_1$ and $\rho_2 = \pi_2$). The marginal survivor functions are $\bar{G}_j(t) = \pi_j^t$, and the sub-survivor functions are

$$\bar{F}(j, t) = \sum_{s=t+1}^{\infty} f(j, s) = \pi_j^t(1-\pi_j)\pi_k^{t+1}/(1-\pi_1\pi_2),$$
$$\bar{F}(\{1, 2\}, t) = \pi_1^t(1-\pi_1)\pi_2^t(1-\pi_2)/(1-\pi_1\pi_2). \tag{5.3.2}$$

The probability of the first failure's occurring in sequence j is

$$p_j = \bar{F}(j, 0) = (1-\pi_j)\pi_k/(1-\pi_1\pi_2),$$

and that of simultaneous failure is

$$p_{\{1,2\}} = \bar{F}(\{1, 2\}, 0) = (1-\pi_1)(1-\pi_2)/(1-\pi_1\pi_2).$$

The marginal distribution of T has survivor function $\bar{F}(t) = \pi_1^t\pi_2^t$ and discrete density $f(t) = \pi_1^{t-1}\pi_2^{t-1}(1-\pi_1\pi_2)$. The marginal hazard functions for this system are $h_j(t) = 1-\pi_j$, the sub-hazard functions are $h(j, t) = (1-\pi_j)\pi_k$ and $h(\{1, 2\}, t) = (1-\pi_1)(1-\pi_2)$, and the overall hazard rate is $h(t) = 1-\pi_1\pi_2$. The $\bar{h}$-functions here are

$$\bar{h}(j, t) = (1-\pi_j)/\pi_j, \quad \bar{h}(\{1, 2\}, t) = (1-\pi_1)(1-\pi_2)/(\pi_1\pi_2), \quad \bar{h}_j(t) = (1-\pi_j)/\pi_j.$$

Note that condition (ii) of Theorem 5.1, which reduces to $\bar{h}(\{1, 2\}, t) = \bar{h}_1(t)\bar{h}_2(t)$ for the case $p = 2$, holds here. This shows that the condition can hold for models with dependent risks, a fact that will be of some significance in Section 7.4. That it is satisfied for this example is predictable from the construction of the model in which the risks act independently up to the first failure. ■

Example 5.3 Discrete version of Marshall-Olkin (1967)

Consider two sequences of Bernoulli trials running in tandem. At any stage, failure might strike either sequence, with probabilities $1-\pi_j$ ($j = 1, 2$), or both together, with probability $1-\pi_{12}$. The three failure mechanisms are assumed to operate independently. Unlike the continuous case, ties can occur here even without dual strikes, i.e., even if $\pi_{12} = 1$.

Let the numbers of failures of the three types over the first t trials be $N_1(t)$, $N_2(t)$, and $N_{12}(t)$. Then the joint survivor function of $\boldsymbol{T} = (T_1, T_2)$ can be derived as

$$\bar{G}(t) = \text{pr}(T_1 > t_1, T_2 > t_2) = \text{pr}\{N_1(t_1) = 0, N_2(t_2) = 0, N_{12}(\max\{t_1, t_2\}) = 0\}$$
$$= \pi_1^{t_1}\pi_2^{t_2}\pi_{12}^{\max(t_1, t_2)}.$$

This has marginals $\bar{G}_j(t_j) = (\pi_j\pi_{12})^{t_j}$ ($j = 1, 2$) and dependence measure (Section 3.1.1) $\bar{R}(t) = \pi_{12}^{-\max(t_1, t_2)}$. The marginal densities for the system are

$$g_j(t) = (\pi_j\pi_{12})^{t-1}(1 - \pi_j\pi_{12}) \quad (j = 1, 2; t = 1, \ldots, m = \infty)$$

and the sub-densities are

$$\begin{aligned} f(j, t) &= \pi_j^{-1}(1-\pi_j)(\pi_1\pi_2\pi_{12})^t \qquad (j = 1, 2), \\ F(\{1, 2\}, t) &= \rho(\pi_1\pi_2\pi_{12})^{t-1}, \end{aligned} \tag{5.3.3}$$

where $\rho = 1 - (\pi_1 + \pi_2 - \pi_1\pi_2)\pi_{12}$ is the probability of failure configuration $\{1, 2\}$ at any given trial. The sub-survivor functions are

$$\begin{aligned} \bar{F}(j, t) &= \pi_j^{-1}(1-\pi_j)(\pi_1\pi_2\pi_{12})^{t+1}/(1-\pi_1\pi_2\pi_{12}), \\ \bar{F}(\{1, 2\}, t) &= \rho(\pi_1\pi_2\pi_{12})^t/(1-\pi_1\pi_2\pi_{12}). \end{aligned} \tag{5.3.4}$$

The probabilities of first failure type are given by

$$\begin{aligned} p_j &= \text{pr}(C = j) = \bar{F}(j, 0) = \pi_j^{-1}(1-\pi_j)(\pi_1\pi_2\pi_{12})/(1-\pi_1\pi_2\pi_{12}), \\ p_{\{1,2\}} &= \text{pr}(C = \{1, 2\}) = \bar{F}(\{1, 2\}, 0) = \rho/(1-\pi_1\pi_2\pi_{12}). \end{aligned}$$

The overall system failure time has density $f(t) = (\pi_1\pi_2\pi_{12})^{t-1}(1 - \pi_1\pi_2\pi_{12})$ and survivor function $\bar{F}(t) = (\pi_1\pi_2\pi_{12})^t$. The marginal hazard functions for this system are $h_j(t) = 1 - \pi_j$, the sub-hazards are $h(j, t) = \pi_j^{-1}(1 - \pi_j)(\pi_1\pi_2\pi_{12})$ and $h(\{1, 2\}, t) = \rho$, and the overall hazard rate is $h(t) = 1 - \pi_1\pi_2\pi_{12}$. The $\bar{h}$-functions here are

$$\bar{h}(j, t) = (1-\pi_j)/\pi_j, \ \bar{h}(\{1, 2\}, t) = \rho/(\pi_1\pi_2\pi_{12}), \ \bar{h}_j(t) = (1-\pi_j\pi_{12})/(\pi_j\pi_{12}).$$

Condition (ii) of Theorem 5.1, which boils down to $\bar{h}(\{1, 2\}, t) = \bar{h}_1(t)\bar{h}_2(t)$ here, evidently fails unless $\pi_{12} = 1$. ■

Example 5.4 Mixture models

Consider p independent sequences of Bernoulli trials with success probabilities π_j ($j = 1, \ldots, p$). The joint survivor function of the p failure times is $\bar{G}(\boldsymbol{t}) = \Pi_{j=1}^p \pi_j^{t_j}$. Now suppose that the sequences are linked through some common random effect $z \in (0, 1)$ such that π_j is replaced by $z\pi_j$. Then the formula just given becomes conditional on z, and the unconditional version is

$$\bar{G}(\boldsymbol{t}) = \Pi_{j=1}^p \pi_j^{t_j} \int_0^1 z^{s_t} dK(z),$$

where $s_t = t_1 + \ldots + t_p$ and K is the distribution function of z. The original formula thus acquires an extra factor, this being the s_tth moment of K.

For example, taking K to be a beta distribution,

$$\begin{aligned}\bar{G}(\boldsymbol{t}) &= \{\Pi_{j=1}^{p}\pi_j^{t_j}\}\int_0^1 z^{s_t}\{z^{\nu-1}(1-z)^{\tau-1}/B(\nu,\tau)\}dz \\ &= \{\Pi_{j=1}^{p}\pi_j^{t_j}\}B(\nu+s_t,\tau)/B(\nu,\tau) \\ &= \{\Pi_{j=1}^{p}\pi_j^{t_j}\}\Pi_{l=1}^{s_t}\{(\nu+s_t-l)/(\nu+\tau+s_t-l)\}.\end{aligned}$$

The sub-density functions for single-index causes of failure are given by

$$\begin{aligned}f(j,t) &= \bar{G}(t,\ldots,t-1,\ldots,t)-\bar{G}(t,\ldots,t) \\ &= \{\Pi_{j=1}^{p}\pi_j\}^t\{\pi_j^{-1}-(\nu+pt-1)/(\nu+\tau+pt-1)\} \\ &\quad \Pi_{l=2}^{pt}\{(\nu+pt-l)/(\nu+pt-l)\},\end{aligned}$$

and those for more complicated configurations can be derived similarly, using Equation 5.2.1, but with rather more trouble. The calculation of the various other associated functions is algebraically untidy though simple to program for computer evaluation. ■

5.4 Likelihood functions

We turn now to estimation, both parametric and non-parametric. Consider the general case where the possible failure times are $\tau_1 < \ldots < \tau_m$, with $\tau_0 = 0$. This makes the transition to continuous failure times in the next chapter smoother, where τ_l becomes an observed t_l. Denote by R_{cl} the set of r_{cl} individuals who fail from cause/configuration c at time τ_l, with $R_l = R_{1l} \cup \ldots \cup R_{pl}$, and R_0 empty, and let $r_l = \sum_c r_{cl}$. Denote by S_l the set of s_l individuals whose failure times are right-censored at time τ_l. For example, in the special but common situation where all individuals are kept under observation up to time τ_a, but none beyond, R_l would be empty for $l > a$ and S_l would be empty for $l \neq a$.

The likelihood function is

$$L = \Pi_{l=1}^{m}\Pi_{i\in R_l}f(c_i,\tau_l;\boldsymbol{x}_i)\times\Pi_{l=0}^{m}\Pi_{i\in S_l}\bar{F}(\tau_l;\boldsymbol{x}_i), \tag{5.4.1}$$

$\boldsymbol{x}_i$ denoting the vector of explanatory variables attached to the ith individual. For an individual whose failure time, but not cause, is observed, a factor $f(\tau_l;\boldsymbol{x})$ must be attached to Equation 5.4.1; likewise, a factor $\bar{F}(c,\tau_l,\boldsymbol{x})$ is needed when the cause is known but the time is (non-informatively) right-censored. We will not explicitly deal with these elaborations here.

In terms of the discrete sub-hazard functions defined in Section 5.1,

$$\bar{F}(\tau_l;\boldsymbol{x}) = \Pi_{s=1}^{l}\{1-h(\tau_s;\boldsymbol{x})\},\ \ f(c,\tau_l;\boldsymbol{x}) = h(c,\tau_l;\boldsymbol{x})\bar{F}(\tau_{l-1};\boldsymbol{x}), \tag{5.4.2}$$

for $l = 1, \ldots, m$, where $h(\tau;x) = \sum_c h(c, \tau;x)$; also, $\bar{F}(\tau_0;x_i) = 1$, so the $l = 0$ term in the second product in Equation 5.4.1 can be dropped. Then, using the same algebraic reduction as from Equation 4.1.7 to Equation 4.1.9,

$$\begin{aligned} L &= \Pi_{l=1}^{m}\Pi_{i\in R_l}[h(c_i, \tau_l;x_i)\Pi_{s=1}^{l-1}\{1-h(\tau_s;x_i)\}] \\ &\times \Pi_{l=1}^{m}\Pi_{i\in S_l}\Pi_{s=1}^{l}\{1-h(\tau_s;x_i)\} \qquad (5.4.3) \\ &= \Pi_{l=1}^{m}[\Pi_{i\in R_l}h(c_i, \tau_l;x_i)\times\Pi_{i\in Q_l}\{1-h(\tau_l;x_i)\}], \end{aligned}$$

where $Q_l = R(\tau_l) - R_l$, $R(\tau_l)$ being the risk set at time τ_l. In this section we will cover the cases (a) fully parametric likelihood, with or without explanatory variables, and (b) non-parametric estimation for random samples without explanatory variables. The case of semi-parametric estimation with explanatory variables will be postponed to Section 6.4.4.

5.4.1 Parametric likelihood

When parametric functions are specified for the discrete sub-densities $f(c, t)$, or equivalently for the sub-hazards $h(c, t)$, L can be used in the standard manner for inference. We will just give one illustration.

Example 5.3 (continued)

Consider the Marshall-Olkin system without covariates. The likelihood function 5.4.1 is

$$L(\pi_1, \pi_2, \pi_{12}) = \Pi_{l=1}^{\infty}\{f_{1l}^{r_{1l}}f_{2l}^{r_{2l}}f_{3l}^{r_{3l}}\}\times\Pi_{l=0}^{\infty}(\pi_1\pi_2\pi_{12})^{l},$$

where, from Equation 5.3.3,

$$\begin{aligned} f_{jl} &= \pi_j^{-1}(1-\pi_j)(\pi_1\pi_2\pi_{12})^{l} \qquad (l = 1, 2) \\ f_{3l} &= \rho(\pi_1\pi_2\pi_{12})^{l-1}, \ \rho = 1-(\pi_1+\pi_2-\pi_1\pi_2)\pi_{12}. \end{aligned}$$

Collecting terms,

$$\begin{aligned} L &= \Pi_{l=1}^{\infty}\{\pi_1^{a_{1l}-r_{1l}}(1-\pi_1)^{r_{1l}}\pi_2^{a_{2l}-r_{2l}}(1-\pi_2)^{r_{2l}}\pi_{12}^{a_{3l}}\rho^{r_{3l}}\} \\ &= \pi_1^{a_{1+}-r_{1+}}(1-\pi_1)^{r_{1+}}\pi_2^{a_{2+}-r_{2+}}(1-\pi_2)^{r_{2+}}\pi_{12}^{a_{3+}}\rho^{r_{3+}}, \end{aligned}$$

where

$$a_{jl} = l(r_l + 1) - r_{3l} \ (j = 1, 2, 3).$$

The log-likelihood derivatives are

$$\begin{aligned}\partial \log L/\partial\pi_1 &= (a_{1+} - r_{1+})/\pi_1 - r_{1+}/(1-\pi_1) - r_{3+}(1-\pi_2)\pi_{12}/\rho,\\ \partial \log L/\partial\pi_2 &= (a_{2+} - r_{2+})/\pi_2 - r_{2+}/(1-\pi_2) - r_{3+}(1-\pi_1)\pi_{12}/\rho,\\ \partial \log L/\partial\pi_{12} &= a_{3+}/\pi_{12} - r_{3+}(\pi_1 + \pi_2 - \pi_1\pi_2)/\rho.\end{aligned}$$

The likelihood equations do not have explicit solution, so iterative maximisation of L is called for. The information matrix can be calculated by further differentiation to provide an estimated covariance matrix for the maximum likelihood estimates of π_1, π_2, and π_{12}. ■

5.4.2 Non-parametric estimation from random samples

For non-parametric inference we follow here Davis and Lawrance's (1989) neat solution of the problem. In the absence of explanatory variables, the likelihood function 5.4.3 is, writing h_{cl} for $h(c, \tau_l)$ and h_l for $h(\tau_l)$,

$$\begin{aligned}L &= \Pi_{l=1}^{m}\Pi_c\{h_{cl}\Pi_{s=1}^{l-1}(1-h_s)\}^{r_{cl}} \times \Pi_{l=1}^{m}\{\Pi_{s=1}^{l}(1-h_s)\}^{s_l}\\ &= \Pi_{l=1}^{m}\{\Pi_c h_{cl}^{r_{cl}} \times (1-h_l)^{q_l - r_l}\},\end{aligned} \tag{5.4.4}$$

where $\Pi_{s=1}^{l-1}(1-h_s)$ is interpreted as 1 for $l = 1$, $r_l = \sum_c r_{cl}$, and

$$q_l = (r_l + s_l) + \ldots + (r_m + s_m)$$

is the size of the risk set $R(\tau_l)$. The estimates $\hat{h}_{cl}$ that maximise L can be calculated by equating

$$\partial \log L/\partial h_{cl} = r_{cl}/h_{cl} - (q_l - r_l)/(1-h_l) \tag{5.4.5}$$

to zero, where $h_l = \Sigma_c h_{cl}$ has been used. This gives

$$\hat{h}_{cl} = r_{cl}(1-\hat{h}_l)/(q_l - r_l),$$

which may be summed over c to yield $\hat{h}_l = r_l/q_l$, and this then leads to

$$\hat{h}_{cl} = r_{cl}/q_l \qquad (l = 1, \ldots, m). \tag{5.4.6}$$

This is the intuitive estimate of h_{cl}, i.e., the observed number of failures divided by the number at risk. If there are no observed failures in the data of type c at time τ_l, the hazard h_{cl} is estimated as zero; this could be true for all c, in which case $r_l = 0$ and $\hat{h}_l = 0$. The maximum likelihood estimates for $\bar{F}(\tau)$ and $f(c, \tau)$ now follow from Equation 5.4.2, and those for the p_c and $\bar{F}(c, \tau)$ (Section 5.1) are

$$\hat{p}_c = \sum_{s=1}^{m} f(c, \tau_s), \hat{\bar{F}}(c, \tau) = \sum \hat{f}(c, \tau_s), \tag{5.4.7}$$

where the second summation is over $\{s:\tau_s > \tau\}$.

There is a slight qualification to the maximum likelihood solution here. In Equation 5.4.4, the mth factor in the likelihood function should be just $\Pi_c h_{cm}^{r_{cm}}$ since $s_m = 0$ implies that $q_m - r_m = 0$ and so $(1 - h_m)^{q_m - r_m} = 1$. As far as the h_{cm} are concerned, then, maximisation of L just entails maximisation of $\Pi_c h_{cm}^{r_{cm}}$ subject to the constraint $\sum_c h_{cm} = h_m = 1$. Luckily, this just yields $\hat{h}_{cm} = r_{cm}/q_m$ as before. However, a real hiccup occurs if q_l becomes zero first at $l = m' \leq m$, which entails $q_l = 0$ for $l = m', \ldots, m$; this will certainly occur if $m = \infty$. The estimates $\hat{h}_{cl} = r_{cl}/q_l$ are then 0/0, and thus undefined for $l \geq m'$, as are the $\hat{h}_l$. An interpretation of this is that, because the conditioning event $\{T > \tau_{l-1}\}$ in the definitions of h_{cl} and h_l is not observed for $l \geq m'$, they are not estimable. On the face of it, then, the estimates of all the other quantities that depend on the $\hat{h}_{cl}$ will also be unobtainable for $l \geq m'$. But all is not lost if $s_{m'-1} = 0$. In that particular case, $r_{m'-1} = q_{m'-1} > 0$, so $\hat{h}_{m'-1} = 1$ which implies $\hat{\bar{F}}(\tau_{m'-1}) = 0$ and then $\hat{\bar{F}}(\tau_l) = 0$ and $\hat{\bar{F}}(c, \tau_l) = 0$ for $l \geq m' - 1$, $\hat{f}(\tau_l) = 0$ and $\hat{f}(c, \tau_l) = 0$ for $l \geq m'$, and $\hat{p}_c = \sum_{s=1}^{m'-1} f(c, \tau_l)$.

When there are no censored observations, the situation is much simpler. In that case, $q_l = r_l + \ldots + r_m$ and, as pointed out in Section 4.2.1, $\hat{\bar{F}}(\tau_l) = q_{l+1}/q_1$, the overall empirical survivor function, q_1 being the sample size. Then, from Equation 5.4.2, $\hat{f}(c, \tau_l) = r_{cl}/q_1$. It follows from Equation 5.4.7 that

$$\hat{\bar{F}}(c, \tau_l) = (r_{c,l+1} + \ldots + r_{cm})/q_1,$$

the sample proportion of type-c failures beyond time τ_l, and then

$$\hat{p}_c = (r_{c1} + \ldots + r_{cm})/q_1,$$

the overall proportion of type-c failures.

Useful plots can be constructed from the estimates. For example, $\hat{\bar{F}}(\tau_l)$ and $\hat{\bar{F}}(c, \tau_l)$ can be plotted against τ_l with either or both scales transformed, e.g., logarithmically, for convenience; plots of the standardised versions, $\hat{\bar{F}}(c, \tau_1)/\hat{p}_c$, are probably better for comparing the shapes of the sub-survivor functions. The hazard functions can be represented in "cusum" form to provide smoother plots, i.e., as $\sum_{s=1}^{l} \hat{h}_s$ and $\sum_{s=1}^{l} \hat{h}_{cs}$ vs. τ_l. The plots will often suggest behaviour such as increasing hazard or failure earlier from one cause than another. Given sufficient data, such hypotheses can be appraised as described later on in this section.

Example 5.5

Let us apply the procedure to data set 1 (King, 1971) of Section 1.2; the breaking strengths are essentially discrete because of the numerical rounding, but there are only failure configurations {1} and {2} present, none of type {1, 2} occurring. For simplicity, we will ignore the suspect values 0 and 3150, and label the distinct, observable breaking strengths as $\tau_1 = 450$, $\tau_2 = 550$, ... The estimates, given in Table 5.1, are apparently based on the assumption, almost certainly false, that $\tau_m = 2050$, i.e., that this is the maximum possible breaking strength. Otherwise, as is quite likely in small samples, the value τ_m was not observed, so we are in danger of the "real hiccup" mentioned above. Fortunately, however, there are no censored values here, so we do not have to hold our breath.

The lay-out in Table 5.1 follows the computations, left to right. Thus, the columns for q_l, r_{1l}, and r_{2l} come directly from inspecting the data; the next three columns, for the hazard functions, are computed from Equation 5.4.6 as $\hat{h}_{cl} = r_{cl}/q_l$ $(c = 1, 2)$ and $\hat{h}_l = (r_{1l} + r_{2l})/q_l$; the next column, for $\hat{\bar{F}}(\tau_l)$, is computed from Equation 5.4.2; the last two columns are computed using $\hat{f}(c, \tau_1)$ from Equation 5.4.2 (carefully noting the subscript $l - 1$) as input to Equation 5.4.7. The estimates for the overall probabilities of the two types of failure, p_1 and p_2, both come out as 0.5 for these data, not surprisingly. Strictly speaking, the rows for $l = 1, 3, 5, 7, 13$, and 14 are redundant in Table 5.1 because, in this small sample, there are no observations at these points.

Plots of the log-survivor functions, $\log\hat{\bar{F}}(\tau_l)$ and $\log\hat{\bar{F}}(j, \tau_l)$ $(j = 1, 2)$ vs. $\tau_l/1000$, are given in Figure 5.1(a); the continuous line is for $\log\hat{\bar{F}}(\tau_l)$, the short-dashed line is for $\log\hat{\bar{F}}(1, \tau_l)$, and the long-dashed line is for $\log\hat{\bar{F}}(2, \tau_l)$. Corresponding plots of the cumulative hazard functions, $\sum_{s=1}^{l}\hat{h}_s$ and $\sum_{s=1}^{l}\hat{h}_{js}$ $(j = 1, 2)$ vs. $\tau_l/1000$, are given in Figure 5.1(b); the points are joined by straight lines just to help the eye though, strictly, these are step functions.

Table 5.1 Estimates of Hazard and Survivor Functions for King (1971) Data

l	τ_l	q_l	r_{1l}	r_{2l}	$\hat{h}_{1l}$	$\hat{h}_{2l}$	$\hat{h}_l$	$\hat{\bar{F}}(\tau_l)$	$\hat{\bar{F}}(1, \tau_l)$	$\hat{\bar{F}}(2, \tau_l)$
1	450	20	0	0	0	0	0	1.00	0.50	0.50
2	550	20	1	0	1/20	0	1/20	0.95	0.45	0.50
3	650	19	0	0	0	0	0	0.95	0.45	0.50
4	750	19	0	1	0	1/19	1/19	0.90	0.45	0.45
5	850	18	0	0	0	0	0	0.90	0.45	0.45
6	950	18	1	1	1/18	1/18	2/18	0.80	0.40	0.40
7	1050	16	0	0	0	0	0	0.80	0.40	0.40
8	1150	16	2	3	2/16	3/16	5/16	0.55	0.30	0.25
9	1250	11	2	0	2/11	0	2/11	0.45	0.20	0.25
10	1350	9	0	1	0	1/9	1/9	0.40	0.20	0.20
11	1450	8	2	1	2/8	1/8	3/8	0.25	0.10	0.15
12	1550	5	1	2	1/5	2/5	3/5	0.10	0.05	0.05
13	1650	2	0	0	0	0	0	0.10	0.05	0.05
14	1750	2	0	0	0	0	0	0.10	0.05	0.05
15	1850	2	0	1	0	1/2	1/2	0.05	0.05	0.00
16	1950	1	0	0	0	0	0	0.05	0.05	0.00
17	2050	1	1	0	1/1	0	1/1	0.00	0.00	0.00

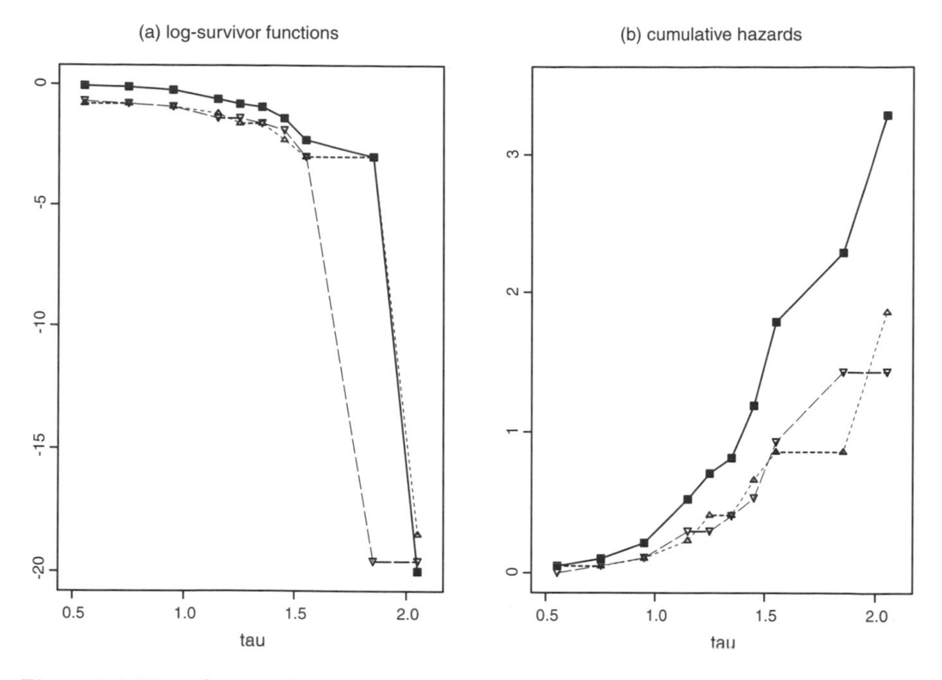

Figure 5.1 King data: (a) log-survivor and (b) cumulative hazard functions.

The two sub-survivor functions look quite similar up to τ_l = 1550: their estimates coincide at several points, notably at time zero in accordance with $\hat{p}_1 = \hat{p}_2 = 0.5$. Equality of $\bar{F}(1, \tau_l)$ and $\bar{F}(2, \tau_l)$ at each τ_l would be equivalent to $h_{1l} = h_{2l}$ ($l = 1, \ldots, m$), i.e., equal hazards from the two causes at each time point. The plots seem to support this conjecture less well at the higher strengths, but the data are rather scanty for definitive conclusions. ■

The procedure described in this section gives estimates of the sub-hazard functions, as in Equation 5.4.6, and thence the sub-densities and sub-survivor functions. In the case of independent risks, more strictly, when the Makeham assumption (Section 3.4.1) holds, these are estimates for the marginal distributions of the components T_j in the latent failure time representation. Zheng and Klein (1995) presented an extended version of this. Independence is replaced by known dependence structure, in the form of a specified copula. Then, a procedure analogous to the Kaplan-Meier method is employed to produce estimates of the marginal distributions.

5.4.3 *Asymptotic distribution of non-parametric estimators*

When m is finite, we can apply standard likelihood theory as in Section 4.1.3 to argue that the mle's $\hat{h}_{cl}$ are jointly asymptotically normal with means h_{cl} and covariances $V_{cl,\,bk} = \text{acov}(\hat{h}_{cl}, \hat{h}_{bk})$ obtained from the information matrix, where acov means "asymptotic covariance." Let $\boldsymbol{h}_l = (h_{1l}, \ldots, h_{dl})^{\mathrm{T}}$ be the $d \times 1$ vector of discrete sub-hazards at time τ_l, where $d = 2^p - 1$, and stack them all into the single $md \times 1$ vector $\boldsymbol{h} = (\boldsymbol{h}_1^{\mathrm{T}}, \ldots, \boldsymbol{h}_m^{\mathrm{T}})^{\mathrm{T}}$. The $V_{cl,\,bk}$ are then found as the elements of the inverse of the $pm \times pm$ matrix $\mathrm{E}(-\partial^2 \log L/\partial \boldsymbol{h}^2)$. Differentiating Equation 5.4.5 with respect to h_{bk}, and using the Kronecker delta symbol,

$$-\partial^2 \log L/\partial h_{cl}\partial h_{bk} = \delta_{lk}\{\delta_{cb} r_{cl}/h_{cl}^2 + (q_l - r_l)/(1 - h_l)^2\};$$

recall that the second term is absent for $l = m$. The presence of δ_{lk} implies that $\hat{h}_l$ and $\hat{h}_k$ are asymptotically independent for $l \neq k$, so the asymptotic covariance matrix $\boldsymbol{V}_h$ of $\hat{h}$ is block-diagonal with m $d \times d$ blocks $\boldsymbol{V}_l = \mathrm{acov}(\hat{h}_l)$ whose elements can be estimated from

$$\begin{aligned}(\hat{V}_l^{-1})_{cb} &= \delta_{cb} r_{cl}/\hat{h}_{cl}^2 + (1 - \delta_{lm})(q_l - r_l)/(1 - \hat{h}_l)^2 \\ &= q_l\{\delta_{cb}/\hat{h}_{cl} + (1 - \delta_{lm})/(1 - \hat{h}_l)\}.\end{aligned}$$

For this asymptotic formula to be useful, we need $\hat{h}_{cl} > 0$ for every c and l. This in turn requires $h_{cl} > 0$ and a sample size large enough to ensure that $r_{cl} > 0$, and so $\hat{h}_{cl} > 0$. In cases where it is known that $h_{cl} = 0$, this parameter would simply be omitted from $\boldsymbol{h}$. In matrix terms,

$$\hat{V}_l^{-1} = q_l \mathrm{diag}\{\hat{h}_{cl}^{-1}\} + q_l(1 - \delta_{lm})(1 - \hat{h}_l)^{-1}\mathbf{J}_d,$$

$\mathbf{J}_d$ being the $d \times d$ matrix of 1s. Thus, inverting the matrix,

$$\hat{V}_{cl,\,bk} = \delta_{lk}(\hat{V}_l)_{cb} = \delta_{lk} q_l^{-1}\hat{h}_{cl}\{\delta_{bc} - (1 - \delta_{lm})\hat{h}_{bl}\}. \tag{5.4.8}$$

For example, it follows from $h_l = \mathbf{1}^{\mathrm{T}}\boldsymbol{h}_l$ that the $\hat{h}_l$ $(l = 1, \ldots, m)$ are asymptotically independent with asymptotic variances estimated as

$$\mathbf{1}_d^{\mathrm{T}}\hat{V}_l^{-1}\mathbf{1}_d = q_l^{-1}\hat{h}_l\{1 - (1 - \delta_{lm})\hat{h}_l\}, \tag{5.4.9}$$

where $\mathbf{1}_d$ is the $d \times 1$ vector of 1s. Likewise, the $\hat{p}_c$ are jointly asymptotically normal with means p_c and covariance matrix calculated as follows. We have, using Equation 5.4.2,

$$p_c = \sum_{l=1}^{m} f(c, \tau_l) = \sum_{l=1}^{m} h_{cl}\bar{F}(\tau_{l-1}) = h_{cl} + \sum_{l=2}^{m}\left\{h_{cl}\sum_{s=1}^{l-1}(1 - h_s)\right\}.$$

Thus, taking differentials,

$$\partial p_c = \partial h_{cl} + \sum_{l=2}^{m}\left[\left\{\partial h_{cl} - h_{cl}\sum_{s=1}^{l-1}\partial h_s/(1 - h_s)\right\}\prod_{s=1}^{l-1}(1 - h_s)\right].$$

Now applying the delta method, the asymptotic covariance between estimates $\hat{p}_c$ and $\hat{p}_b$ may be evaluated as

$$\begin{aligned} E\{(\partial\hat{p}_c)(\partial\hat{p}_b)\} \sim\ & V_{c1,b1} - \sum_{l=2}^{m} h_{bl}S_{l-1}V_{cl,+1}/(1-h_1) + \sum_{l=2}^{m} S_{l-1}^2 V_{cl,bl} \\ & - \sum_{l=2}^{m}\sum_{l'>l}(h_{bl'}V_{cl,+1} + h_{cl'}V_{bl,+1})S_{l-1}S_{l'-1}/(1-h_l) \\ & - \sum_{l=2}^{m}\sum_{l'=2}^{m} h_{cl}h_{bl'}S_{l-1}S_{l'-1}\sum_{s=2}^{l''-1}\left(\sum_{j=1}^{p} V_{js,+s}\right)/(1-h_s)^2 \end{aligned}$$

where we have used $h_s = h_{1s} + \ldots + h_{ps}$ and $E\{(\partial\hat{h}_{cl})(\partial\hat{h}_{bk})\} \sim V_{cl,bk}$, which is zero for $l \neq k$, and adopted the notations S_l for $\Pi_{s=1}^{l}(1-h_s)$, $V_{cl,+l}$ for $\sum_{j=1}^{p} V_{cl,jl}$ and l'' for min(l, l'). This expression for $\text{acov}(\hat{p}_c, \hat{p}_b)$ is not a pretty sight but is easily computed.

Linear hypotheses involving the h_{cl} can be tackled as follows. Consider H:$\boldsymbol{Ah} = \boldsymbol{a}$, where $\boldsymbol{A}$ is a given $r \times md$ matrix of full rank r, and $\boldsymbol{a}$ is a given $r \times 1$ vector. A routine test for H can be based on the Wald statistic

$$W = (\boldsymbol{A}\hat{h} - \boldsymbol{a})^{\mathrm{T}}(\boldsymbol{A}\boldsymbol{V}_h\boldsymbol{A}^{\mathrm{T}})^{-1}(\boldsymbol{A}\hat{h} - \boldsymbol{a});$$

under H, W is asymptotically distributed as χ_r^2.

Example 5.5 (continued)

We will apply the asymptotic formulae to examine a couple of hypotheses concerning the h_{cl}, though this can only be a pure illustration with such a small sample.

Consider first H_0:$h_{1l} = h_{2l}$ ($l = 1, \ldots, m$) suggested by the plots in Figure 5.1. To construct the Wald test statistic for this hypothesis we take $\boldsymbol{a} = \boldsymbol{0}$ and $\boldsymbol{A}$ as the $m \times 2m$ matrix whose jth row has 1 in position $2j-1$, -1 in position $2j$, and zeros elsewhere. From Equation 5.4.8, $\boldsymbol{V}_l$, and therefore $\boldsymbol{V}_h$, is singular if $\hat{h}_{cl}$ is zero for some c. However, provided only that the τ_l with $\hat{h}_l = 0$ are omitted, it is easy to see that $\boldsymbol{A}\boldsymbol{V}_h\boldsymbol{A}^{\mathrm{T}}$ is non-singular and so the Wald statistic is computable. Referring to Table 5.1, we have to delete seven rows, leaving $m = 10$, because of the limited sample size. The numerical result is $W = 9.47$ as χ_{10}^2 (p-value > 0.10).

As a second illustration, consider H_0:$h_l = h_1$ ($l = 1, \ldots, m$); H_0 here stipulates a constant hazard as in the geometric structure of Example 5.1. For the Wald statistic take $\boldsymbol{a} = 0$ and $\boldsymbol{A}$ as the $(m-1) \times 2m$ matrix whose jth row is (1, 1, 0… 0, –1, –1, 0 … 0), the –1s being in positions $2j + 1$ and $2j + 2$. The computation yields $W = 15.52$ as χ_9^2 ($0.05 < p$-value < 0.10).

chapter 6

Hazard-based methods for continuous failure times

6.1 Latent failure times vs. hazard modelling

As indicated in Chapter 3, the traditional approach to modelling Competing Risks is via a multivariate latent failure time distribution $\bar{G}(\boldsymbol{t})$, moreover, one in which the T_j are independent. The modern approach is more hazard-conscious: the framework for modelling and inference is based on the sub-hazard functions.

Prentice et al. (1978), reiterated in Kalbfleish and Prentice (1980, Section 7.1.2), argued strongly against the traditional approach. Their attack was based partly on the non-identifiability of $\bar{G}(\boldsymbol{t})$ (Theorem 7.1 below). They also pointed out that the traditional aim of isolated inference about a subset of components of $\boldsymbol{T}$ makes unjustifiable assumptions about the effect of removing the other risks, even if this were possible. Finally, they cast doubts on the physical existence of hypothetical latent failure times in many contexts. These points have been discussed also in Section 3.3. Thus, one should set up models only for observable phenomena, a kind of "What you see is what you set" doctrine.

The important arguments of Prentice et al. (1978) deserve examination. The lack of identifiability is softened to some extent in the regression case (Theorem 7.6). That removal of some risks might change the circumstances applies equally to hazard models, in fact to any circumstance in Statistics where one is trying to extrapolate beyond what has been observed. Finally, the one centering upon the "lack of physical meaning" of latent failure times seems to be arguable in some cases. A case for the defence, of using a multivariate failure time distribution, has been given by Crowder (1994). Other support for such heresy is given by Aalen (1995): "Maybe the models should go further than the data, and that the amount of speculation this entails may be fruitful."

In chewing over the arguments, what is perhaps the main recommendation for the hazard-based approach has not yet been mentioned. This is that models for processes evolving over time can be developed much more naturally in terms of hazards than multivariate survivor functions. Thus, one can deal with quite complex situations that would be difficult, even intractable, from the traditional point of view. A vast amount of modern methodology and interpretation has now been developed and implemented around hazard functions.

6.2 Some examples of hazard modelling

The usual broad classification of hazard functions is into three types: (i) constant over time, as would be appropriate in a constant environment subject only to random shocks; (ii) increasing, as in the case of a steadily deteriorating system; (iii) decreasing, as for systems which wear in and become less liable to failure as time goes on. The univariate Weibull distribution is the usual start for discussing different types of behaviour of hazard functions, and we shall honour tradition. Thus, $h(t) = \phi\xi t^{\phi-1}$ is the hazard function corresponding to the Weibull survivor function $\bar{F}(t) = \exp(-\xi t^{\phi})$. It is (i) constant for $\phi = 1$, corresponding to an exponential distribution, (ii) increasing for $\phi > 1$, and (iii) decreasing for $\phi < 1$.

The construction of customized hazard functions, designed to suit the application at hand, has received some attention. Various methods have been proposed, and we shall just look at a very straightforward one.

Example 6.1 Bathtub hazard function

To simulate this legendary shape (down-along-up) we might simply add together the three Weibull possibilities, i.e., take the hazard function as

$$h(t) = \xi_0 + \phi_1\xi_1 t^{\phi_1 - 1} + \phi_2\xi_2 t^{\phi_2 - 1},$$

where $\phi_1 > 1$ and $\phi_2 < 1$. The corresponding survivor function is

$$\bar{F}(t) = \exp\left\{-\int_0^t h(s)ds\right\} = \exp\{-\xi_0 t - \xi_1 t^{\phi_1} - \xi_2 t^{\phi_2}\} = \bar{G}_0(t)\bar{G}_1(t)\bar{G}_2(t),$$

where the $\bar{G}_j$ ($j = 0, 1, 2$) are the survivor functions of the three contributing distributions. The corresponding representation in terms of random variables is $T = \min\{T_0, T_1, T_2\}$ where the T_j are independent with survivor functions $\bar{G}_j$. In Reliability applications, the bathtub hazard is relevant when there is relatively high risk of early failure ("wear in") and of late failure ("wear out"). In medical applications, the bathtub hazard can reflect treatment with non-negligible operative mortality which is otherwise life-preserving; as Boag (1949) put it, it is not the malady but the remedy which can prove fatal. ■

The construction illustrated in Example 6.1, whereby a hazard form is constructed by combining several contributions, is generally applicable, not just confined to the minimum of three independent Weibull variates.

In standard Survival Analysis, we have to consider one hazard function, and this causes enough problems. In the Competing Risks context, we have to consider simultaneously a whole set of them, the $h(j, t)$ for $j = 1, \ldots, p$. In previous chapters we have looked at the forms of hazard functions implied by certain joint latent failure time distributions (e.g., Examples 3.2 and 3.3). We now consider one or two others from the literature.

Example 6.2 Exponential mixture (Example 1.2)

The sub-survivor functions are $\bar{F}(j, t) = \pi_j e^{-\lambda_j t}$ and the sub-hazards are

$$h(j, t) = \pi_j \lambda_j e^{-\lambda_j t} / \sum_k \pi_k e^{-\lambda_k t}.$$

Let $\lambda_m = \min(\lambda_1, \ldots, \lambda_p)$, then $\int_0^\infty h(m, s)ds < \infty$ and this makes the mth component of the independent-risks proxy distribution improper, i.e., $\bar{F}^*_m(\infty) > 0$. Now, none of this matters in the least unless the situation calls for an interpretation in terms of latent failure times. Then, if the risks were actually independent, or even if only the partial Makeham condition $h_m(t) = h(m, t)$ held, we would be forcing $\bar{F}_m(\infty) > 0$, and so unwittingly assuming T_m to be an improper failure time, i.e., $\text{pr}(T_m = \infty) > 0$. In contrast, with independent exponential latent lifetimes, $\bar{F}_j(t) = e^{-\lambda_j t}$ and $h(j, t) = h_j(t) = \lambda_j$. ■

That this kind of accident can happen was pointed out by Nadas (1970) in the context of independent risks. We do not have an entirely free hand in constructing models for the sub-distributions. If there is to be a latent lifetime structure in the background, then there are dangers in not deriving the sub-distributions from an underlying joint latent failure time distribution.

Example 6.3 Bivariate exponential (Gumbel, 1960; Example 3.1)

The sub-hazard functions for this distribution are $h(j, t) = \lambda_j + \nu t$. In terms of the broad classification made above, this represents the sum of a constant hazard, λ_j, and an increasing one, νt, remembering that $\nu > 0$. This form might be appropriate for a system in which the jth component both deteriorates and is subject to random shocks over time. ■

Example 6.4 Bivariate Makeham (Arnold and Brockett, 1983)

The univariate Gompertz distribution has hazard function $\psi e^{\phi t}$, with $\phi > 0$ and $\psi > 0$. This represents an exponential rate of deterioration of a system. If we add the possibility of pure chance failure, represented by a constant hazard λ, we obtain the univariate Makeham survivor function

$$\exp\left\{-\int_0^t (\lambda + \psi e^{\phi s}) ds\right\} = \exp\{-\lambda t + \psi\phi^{-1}(1 - e^{\phi t})\}.$$

Arnold and Brockett combined this with the Marshall-Olkin type of simultaneous failure (Section 7.2, below) to produce a bivariate survivor function of form

$$\bar{G}(\boldsymbol{t}) = \exp\{-\lambda_1 t_1 - \lambda_2 t_2 - \lambda_{12}\max(t_1, t_2) - \psi_1\phi_1^{-1}(1 - e^{\phi_1 t_1}) - \psi_2\phi_2^{-1}(1 - e^{\phi_2 t_2})\}.$$

This is an example of customizing hazard functions: the sub-hazards,

$$h(j, t) = \lambda_j + \psi_j e^{\phi_j t} \ (j = 1, 2), \ h(\{1, 2\}, t) = \lambda_{12},$$

are constructed to have the intended functional behaviour. Here T can be expressed as $\min(R_1, R_2, R_{12}, W_1, W_2)$ where the Rs are independent, exponentially distributed times to failures of types {1}, {2}, and {1, 2}, respectively, and W_1 and W_2 are Gompertz-distributed. Arnold and Brockett showed that the bivariate latent failure time distribution is parametrically identifiable by the sub-hazard functions, i.e., by Competing Risks data. ■

Example 6.5 Kimber and Grace: the dream team

Many of the published applications of Competing Risks concern death and disease among humans and animals. To strike a more cheerful note, let us look at some cricketing data.

The first data set has been sportingly provided for our entertainment by my colleague, Alan Kimber. Alan played cricket at a high level for a good number of years (somewhat higher than my own level in football). His paper (Kimber and Hanson, 1993) applied techniques of Survival Analysis to batting scores in cricket. The failure time is taken as the number of runs scored until "out," right-censored if "not out." The runs may be regarded as a proxy for the time at the wicket or as reflecting better than clock time the wear and tear on the batsman. (This might account for Kimber's ever-youthful appearance.) Hazard modelling is natural in this context: each time the batsman faces the bowler, there is a clear risk of mistake. In fact, it is cricketing folklore that the hazard varies critically with runs scored, e.g., there may be an increased risk early on, until the batsman gets his "eye in" and, if he gets that far, he runs into the "nervous nineties," when the anxiety to reach a century dominates. If, besides the number of runs, we take note of the cause of failure, i.e., how the batsman was dismissed, the data become of Competing Risks type. The officially admitted causes of failure here are b = bowled, c = caught, l = lbw (leg before wicket), s = stumped, r = run out, and n = not out. Table 6.1 gives Alan Kimber's batting record during his time playing

Table 6.1 Kimber's Batting Data for Guildford: (t, c) = (Runs, How Out)

1981	2 b	0 b	22 c	120 b	61 r	5 b	32 n	24 b	27 r	
1982	55 n	84 b	43 b	67 c	6 c	49 s	0 c	13 b	12 c	39 c
	7 c	51 c	1 b	18 l	16 c	10 c	12 c	11 b	2 c	30 n
	42 c	77 c	53 b	17 r	34 n	35 n	6 c	40 n	41 n	
1983	32 n	16 b	28 b	36 b	20 c	1 r	3 n	95 n	3 n	23 c
	10 b	5 s	56 n	0 c	11 b	9 c	22 b	12 c	34 b	40 n
	41 n									
1984	19 c	25 c	1 c	20 l	29 c	41 b	0 n	40 n	0 n	4 n
	1 n	20 n	46 c	11 c	35 c	23 c	39 b	17 c	42 n	
1985	12 s	15 l	0 l	6 c	31 n	4 l	1 n	23 c	0 n	6 n
	11 b	19 l	9 c	41 c	0 r	31 c				
1986	13 l	43 b	11 l	25 c	7 n	43 c	8 n	21 r	4 b	42 c
	6 n	27 n	29 c	1 b						
1987	11 n	28 n	37 n	6 b	51 n	17 l	0 b	3 c	0 c	14n
	0 b	25 c	10 l	7 c	3 c					
1988	11 c	58 c	2 b	36 c						
1989	42 n	9 c	5 c	2 c	13 c	12 b	52 n	21 b		
1990	25 c	8 b	100 n	6 c	13 c	14 c	48 n	23 c	12 c	

for Guildford, 1981 to 1990. (He maintains that his century in 1990 would have been augmented but for the intervention of tea!)

Models with constant hazard, and with increased hazard when the number of runs is less than 10 or in the 90s, were fitted. Without wishing to be in any way judgemental, it has to be said in a caring way that there is not enough data in the 90s to address that particular anxiety. The appropriate likelihood function for discrete failure times is given in Equation 5.4.1: here, τ_l can take values 0, 1, 2, ..., $\bar{F}(\tau_l;x_i)$ is expressed as $\Pi_{s=0}^{l}\{1 - h(\tau_l)\}$ and $f(c_i, \tau_l;x_i)$ as $h(c_i, \tau_l)\Pi_{s=0}^{l-1}\{1 - h(\tau_l)\}$, there being no explanatory variables. In turn, for the constant-hazard model $h(c, \tau_l)$ is expressed as $h_c = \exp(-\theta_c)/\{1 + \sum_{j=1}^{5}\exp(-\theta_j)\}$, for $c = 1, \ldots, 5$, the θ_c being parameters of unrestricted range (Appendix 1). The resulting log-likelihood is –593.68607. For the modified hazards model, h_c is replaced by $h_c\exp(\theta_6)$ whenever τ_l was below 10 or in the 90s, the exponential factor representing the anxiety effect. The log-likelihood for this model is 593.67800, giving log-likelihood ratio chi-square $\chi_1^2 = 0.016$ ($p \approx 0.90$), the estimated anxiety factor actually being very slightly less than 1.0. Either Kimber has nerves of steel or he doesn't know what the score is. Just for interest, the sub-hazard estimates come out as 0.009, 0.017, 0.003, 0.001, 0.002, with standard errors suggesting that no two are equal; their sum, the overall hazard, is 0.032. The sub-hazard for cause 2 is as large as the rest put together, illustrating the batsman's typical generosity in always being ready to give a catch to the fielders.

Another grand old man of English cricket was Dr. W. G. Grace. His scores from matches in the latter half of the 19th century are to be found in Lodge (1990). In Grace's case, possibly because the data are more extensive than Kimber's, the anxiety factor does show moderate significance ($\chi_1^2 = 3.90$, $p < 0.05$), its estimate being 1.35; the sub-hazard estimates show that WGG's preferred mode of departure was to be caught out, being bowled out coming a fairly close second. ■

6.3 Non-parametric methods for random samples

Methods similar to those outlined in Chapter 4 for Survival Analysis can be extended to cope with Competing Risks.

6.3.1 The Kaplan-Meier estimator

We begin by extending the Kaplan-Meier estimator (Section 4.2), following Davis and Lawrance (1989) as in Section 5.5.2.

Suppose that the observed failure times, from all causes, are $t_1 < t_2 < \ldots < t_m$, and take $t_0 = 0$ and $t_{m+1} = \infty$. Denote by R_{jl} the set of r_{jl} individuals who fail from cause j at time t_l; then r_{jl} is zero and R_{jl} is empty for $l = 0$ and $l > m$. Also, unless there are tied failures from different causes at t_l, only one of $r_{1l}, \ldots, r_{pl}$ will be non-zero. Let $R_l = R_{1l} \cup \ldots \cup R_{pl}$ and denote by S_l the set of s_l individuals whose failure times t_{ls} $(s = 1, \ldots, s_l)$ are right-censored during $[t_l, t_{l+1})$.

We take the form of the likelihood function as in Equation 5.5.1, but without explanatory variables and with t_l in place of τ_l:

$$L = \Pi_{l=1}^{m} \Pi_{i \in R_l} \Delta\bar{F}(c_i, t_l) \times \Pi_{l=0}^{m} \Pi_{i \in S_l} \bar{F}(t_{ls}), \tag{6.3.1}$$

where

$$\Delta\bar{F}(c_i, t_l) = \bar{F}(c_i, t_l-) - \bar{F}(c_i, t_l).$$

Now, the maximum likelihood estimate of $\Delta\bar{F}(c_i, t_l)$ cannot be zero, since this would make L zero, so $\bar{F}(c_i, t_l)$ must be discontinuous at t_l whenever $i \in R_l$. Also, for L to be maximised it is necessary to take $\hat{\bar{F}}(t_{ls}) = \hat{\bar{F}}(t_l)$ for each individual $i \in S_l$ (as in Section 4.2). For this, one must take $\hat{\bar{F}}(j, t_{ls}) = \hat{\bar{F}}(j, t_l)$ for each j; this is because $\bar{F}(t) = \sum_j \bar{F}(j, t)$ and each $\bar{F}(j, t)$ is monotone decreasing in t. In particular, $\hat{\bar{F}}(t_{0s}) = \hat{\bar{F}}(t_0) = 1$, so the second product term in Equation 6.3.1 effectively begins at $l = 1$ rather than at $l = 0$. Hence, we have to maximize

$$\Pi_{l=1}^{m} [\Pi_{j=1}^{p} \{\Delta\bar{F}(j, t_l)\}^{r_{jl}} \times \{\bar{F}(t_l)\}^{s_l}],$$

where $\bar{F}(j, t)$ has a discontinuity at t_l if $r_{jl} > 0$. In terms of the corresponding discrete sub-hazard functions (Section 5.1),

$$\Delta\bar{F}(j, t_l) = h(j, t_l)\bar{F}(t_{l-1}), \quad \bar{F}(t_l) = \Pi_{s=1}^{l}\{1 - h(t_s)\}, \tag{6.3.2}$$

where $h(t) = \sum_j h(j, t)$, and so the $h(j, t_l)$ have to be found to maximise

$$\Pi_{l=1}^{m}\{\Pi_{j=1}^{p}[h(j, t_l)\Pi_{s=1}^{l-1}\{1-h(t_s)\}]^{r_{jl}} \times [\Pi_{s=1}^{l}\{1-h(t_s)\}]^{s_l}\}$$
$$= \Pi_{l=1}^{m}[\Pi_{j=1}^{p} h(j, t_l)^{r_{jl}} \times \{1-h(t_l)\}^{q_l - r_l}]$$

where $r_l = r_{1l} + \ldots + r_{pl}$ and

$$q_l = (r_l + s_l) + \ldots + (r_m + s_m)$$

as in Equation 5.5.4. The maximum likelihood estimates are

$$\hat{h}(j, t_l) = r_{jl}/q_l, \; \hat{h}(t_l) = r_l/q_l, \tag{6.3.3}$$

and those for $\bar{F}(t)$, p_c, and $\bar{F}(j, t)$ follow as described in Section 5.4.2. Since $q_m > 0$ necessarily, the hiccup referred to in Section 5.5.2 is cured. However, if $s_m > 0$, $\hat{h}(t_m) < 1$ and so $\hat{\bar{F}}(t_m) > 0$. Then, as in Section 4.2, $\hat{\bar{F}}(t)$ is undefined for $t > t_m$, which implies the same for all the other estimates.

Plots of the estimated functions can be made as described in Section 5.5.2. Note the distinction between cumulative and integrated hazard functions (Section 4.2.2). In the case of continuous failure times, $-\log\bar{F}(t)$ is the integrated overall hazard function, not the cumulative one. Again, $-\log\bar{F}(j, t)$ is neither the integrated nor the cumulative sub-hazard function.

Example 6.6

The estimates are derived here for Hoel's (1972) data, data set 4 of Section 1.2. Some details of the computations are given in Table 6.2 for Group 1 of the data; the survival times have been scaled down by a factor of 1000. The sample size is 99, and there are 4 tied times, so the full table has 95 rows of which only the first 5 and last 5 are shown here. The computations are essentially the same as those for Example 5.4 shown in Table 5.1.

Plots of the log-survivor and cumulative hazard functions are shown in Figure 6.1 for Groups 1 and 2 separately. In each case, the overall log-survivor curve is at the top, since the overall survivor function is larger than each sub-survivor function, being their sum. Also in each case, the cause-1 curve is at the bottom, though only for times beyond about 0.25 for Group 2. We might be tempted to conclude from this that the cause-1 events tend to occur earlier than the others. If so, we would have fallen nicely into a trap that others would be keen to point out, purely in the interests of scientific truth, of course. Recall that $\bar{F}(c, t)$ is pr($C = c$, $T > t$), which is not the survivor function for type-c failures; the latter is a conditional probability, $\bar{F}(c, t)/p_c$, where $p_c = \text{pr}(C = c) = \bar{F}(c, 0)$ (Section 1.1.2). Thus, to arrive at the conclusion stated, we first need to standardise the survivor functions, by dividing by the corresponding p_c, so that the

Table 6.2 Estimates of Hazard and Survivor Functions for Hoel (1972) Data, Group 1

l	t_l	$\hat{h}_{1l}$	$\hat{h}_{2l}$	$\hat{h}_{3l}$	$\hat{h}_l$	$\hat{\bar{F}}(1, t_l)$	$\hat{\bar{F}}(2, t_l)$	$\hat{\bar{F}}(3, t_l)$	$\hat{\bar{F}}(t_l)$
1	0.040	0.000	0.000	0.010	0.010	0.222	0.384	0.384	0.990
2	0.042	0.000	0.000	0.010	0.010	0.222	0.384	0.374	0.980
3	0.051	0.000	0.000	0.010	0.010	0.222	0.384	0.364	0.970
4	0.062	0.000	0.000	0.010	0.010	0.222	0.384	0.354	0.960
5	0.159	0.011	0.000	0.000	0.011	0.212	0.384	0.354	0.949
...	...	...	...	...	...	...	...	...	...
91	0.738	0.000	0.200	0.000	0.200	0.000	0.020	0.020	0.040
92	0.748	0.000	0.250	0.000	0.250	0.000	0.010	0.020	0.030
93	0.753	0.000	0.333	0.000	0.333	0.000	0.000	0.020	0.020
94	0.761	0.000	0.000	0.500	0.500	0.000	0.000	0.010	0.010
95	0.763	0.000	0.000	1.000	1.000	0.000	0.000	0.000	0.000

standardised versions all start at value 1 at time 0. In Figure 6.1 this just means shifting them vertically to start at 0. It is clear that when this is done the stated conclusion does, in fact, stand. Regarding failure-types 2 and 3, these are represented by the middle curves, with type-2 above type-3 in Group 1 and vice versa in Group 2. After the standardising shift, the relative positions are unchanged in Group 1, but the curves converge in Group 2 for times up to about 0.7, after which the type-2 curve drops more steeply. The picture does not admit a simple, one-line description.

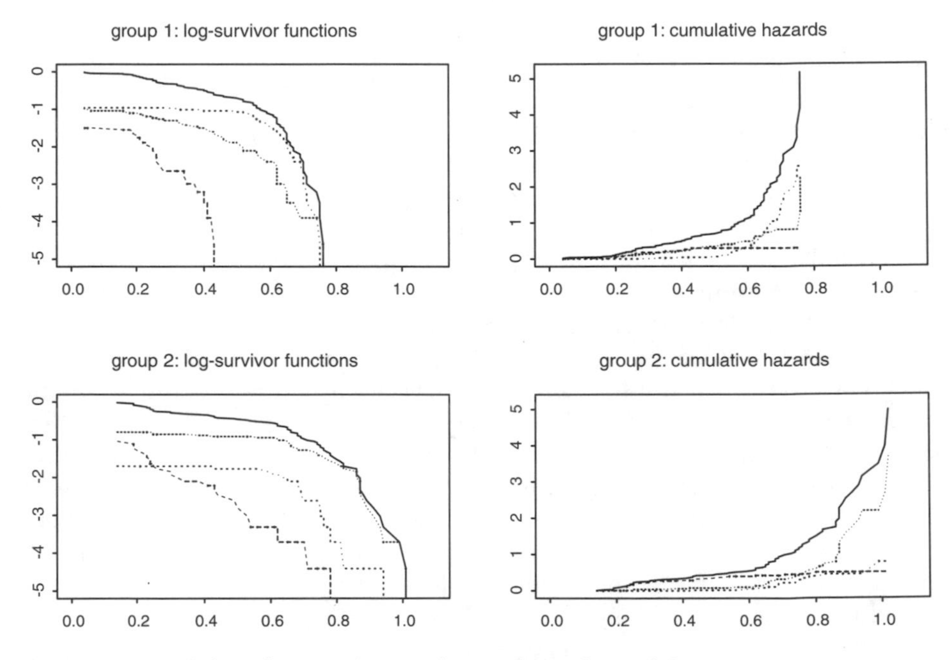

Figure 6.1 Hoel data: log-survivor and cumulative hazard functions.

The Weibull-mixture model, fitted to these data in Example 2.5, can accommodate log-sub-survivor functions at different initial levels and of different shapes, as here, provided that the general form of each curve is a constant plus a power of t. However, it will be recalled that it was not found possible there either to achieve a simpler picture by reducing the full parameter set. ■

Dinse (1982) extended the likelihood Equation 6.3.1 to accommodate cases with known failure times and unknown failure causes and cases with known failure causes but non-informatively right-censored failure times; the corresponding likelihood contributions are, respectively, $\Delta\bar{F}(t)$ and $\bar{F}(c, t)$. Dinse presented an iterative numerical method for computing the non-parametric maximum likelihood estimates in this situation.

Pepe (1991) and Pepe and Mori (1993) proposed other summary curves for describing competing risks data: they recommended the use of functions associated with cumulative incidence, prevalence, and marginal and conditional probabilities and gave methodology for estimating and testing these functions.

6.3.2 *Interval-censored data*

The development in Section 5.5, though for discrete failure times, applies almost immediately to interval-censored, or grouped, continuous failure times. Suppose that the grouping intervals are $I_l = (\tau_{l-1}, \tau_l]$, for $l = 1, \ldots, m$. The appropriate likelihood function is Equation 5.5.1, with $f(c_i, \tau_l; x_i)$ replaced by $\bar{F}(c_i, \tau_{l-1}; x_i) - \bar{F}(c_i, \tau_l; x_i)$. The equivalent form Equation 5.5.3, in terms of hazard contributions, follows.

In the case of random samples, without explanatory variables, we have, from Equation 5.5.6, the hazard estimates $\hat{h}_{jl} = r_{jl}/q_l$ and $\hat{h}_l = r_l/q_l$ for

$$h_{jl} = \text{pr(failure from cause } j \text{ in } I_l \mid \text{enter } I_l)$$

and

$$h_l = \text{pr(failure in } I_l \mid \text{enter } I_l).$$

Then pr(survive to enter I_l) and pr(failure from cause j in I_l) are estimated from Equation 5.5.2 as

$$\hat{\bar{F}}(\tau_{l-1}) = \Pi_{s=0}^{l-1}(1 - r_s/q_s) \quad \text{and} \quad f(j, \tau_l) = \hat{h}_{jl}\hat{\bar{F}}(\tau_{l-1}),$$

and then, from Equation 5.5.2, pr(eventual failure from cause j) as $\hat{p}_j = \sum_{l=1}^{m} \hat{f}(j, \tau_l)$.

Example 6.7

Data set 6 of Section 1.2 (Nelson, 1982) provides an example where the failure times, numbers of cycles to failure in fatigue testing of a superalloy, have been grouped on a log-scale. The τ_l-values, together with the estimates of the hazard and survivor functions, are given in Table 6.3.

Plots of the survivor functions, log-transformed, and of the cumulative hazards are given in Figure 6.2. The latter show a fairly clear pattern: the overall cumulative hazard is the top curve, being the sum of the other two, and the cumulative sub-hazard for defect-type 1 does not really get off the ground until about $\tau = 4.05$ after which it rises more sharply, eventually overtaking the sub-hazard for defect-type 2, which has flattened out. It seems, then, that defects of type 2 are predominant at lower numbers of cycles, and then defects of type 1 catch up and overtake later on, at which stage defects of type 2 have all but disappeared. The log-survivor functions tell a similar tale (the same tale, strictly speaking): the overall curve is at the

Table 6.3 Estimates of Hazard and Survivor Functions for Nelson (1982) Data

l	τ_l	$\hat{h}_{1l}$	$\hat{h}_{2l}$	$\hat{h}_l$	$\hat{\bar{F}}(1, \tau_l)$	$\hat{\bar{F}}(2, \tau_l)$	$\hat{\bar{F}}(\tau_l)$
1	3.55	0.000	0.008	0.008	0.156	0.836	0.992
2	3.60	0.000	0.016	0.016	0.156	0.820	0.977
3	3.65	0.000	0.048	0.048	0.156	0.773	0.930
4	3.70	0.003	0.064	0.067	0.154	0.714	0.867
5	3.75	0.009	0.141	0.150	0.146	0.591	0.737
6	3.80	0.004	0.155	0.159	0.143	0.477	0.620
7	3.85	0.000	0.239	0.239	0.143	0.328	0.471
8	3.90	0.006	0.204	0.210	0.141	0.232	0.372
9	3.95	0.000	0.140	0.140	0.141	0.180	0.320
10	4.00	0.000	0.130	0.130	0.141	0.138	0.279
11	4.05	0.028	0.112	0.140	0.133	0.107	0.240
12	4.10	0.033	0.043	0.076	0.125	0.096	0.221
13	4.15	0.024	0.035	0.059	0.120	0.089	0.208
14	4.20	0.062	0.125	0.188	0.107	0.063	0.169
15	4.25	0.031	0.046	0.077	0.102	0.055	0.156
16	4.30	0.050	0.067	0.117	0.094	0.044	0.138
17	4.35	0.132	0.132	0.264	0.076	0.026	0.102
18	4.40	0.308	0.103	0.410	0.044	0.016	0.060
19	4.45	0.304	0.087	0.391	0.026	0.010	0.036
20	4.50	0.286	0.214	0.500	0.016	0.003	0.018
21	4.55	0.286	0.143	0.429	0.010	0.000	0.010
22	4.60	0.250	0.000	0.250	0.008	0.000	0.008
23	4.65	0.333	0.000	0.333	0.005	0.000	0.005
24	4.70	0.000	0.000	0.000	0.005	0.000	0.005
25	4.75	0.500	0.000	0.500	0.003	0.000	0.003
26	4.80	0.000	0.000	0.000	0.003	0.000	0.003
27	4.85	1.000	0.000	1.000	0.000	0.000	0.000

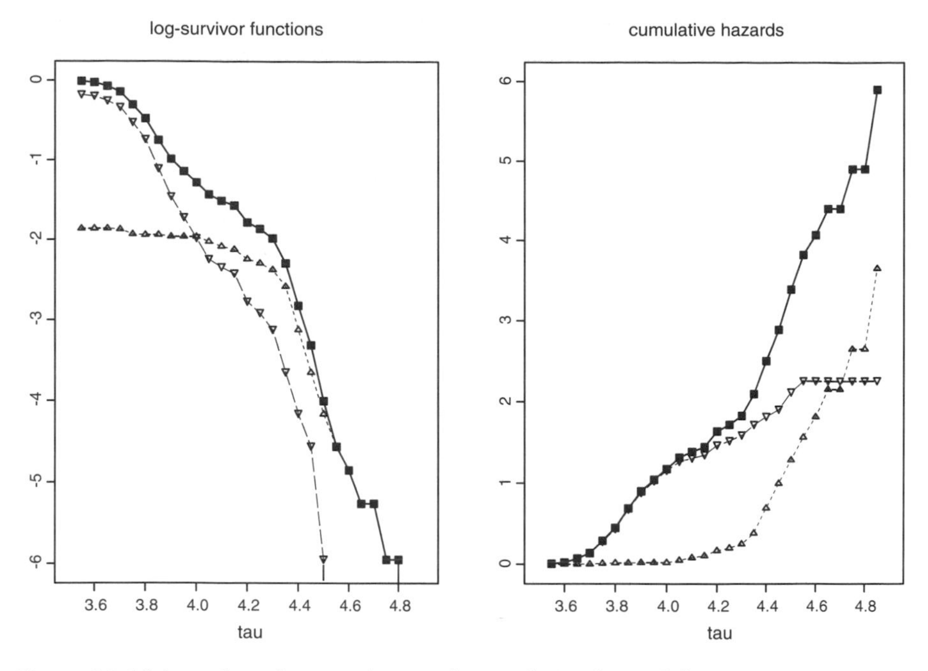

Figure 6.2 Nelson data: log-survivor and cumulative hazard functions.

top with the defect-2 curve in close attendance at the lower numbers of cycles; the defect-1 curve starts off lower, but is flatter and crosses its rival at about $\tau = 4.05$. ■

6.3.3 *Actuarial approach*

The h_{jl} and h_l are probabilities of observable events and are thus estimable without too much fuss. In actuarial and demographic work, calculations have routinely been made of certain unobservable aspects from observed ones. Thus, estimates have been derived of the probabilities of death from one set of causes with another set of risks having been eliminated in some way. Putting aside the question of the real-life relevance of such calculations (Section 3.3), it is inevitable that some assumption connecting the observable and unobservable aspects has to be made. In different contexts such assumptions may have greater or lesser credibility. The Makeham assumption is an example: it says that the net (unobservable) hazard $h_j(t)$ is equal to the crude (observable) one, $h(j, t)$. We know from Theorem 3.3 that under this assumption the marginals $\bar{G}_j(t)$ are in principle determined by the sub-survivor functions $\bar{F}(j, t)$. A variety of such assumptions is recorded in the actuarial and demographic literature. In the remainder of this section we just give a brief glimpse of the field, more extensive treatments being available in Chiang (1968), Gail (1975), David and Moeschberger (1978), and the references therein.

Let us consider some typical assumptions and the estimation of some unobservable aspects. The assumptions, involving the continuous-time hazard functions, are

(i) that of Makeham, $h_j(t) = h(j, t)$ for each j and all $t > 0$;
(ii) that of Chiang (1961b), the proportionality $h_j(t) = \omega_{jl}h(t)$ for $t \in I_l$;
(iii) that of Kimball (1969), that the effect of eliminating risk j is to modify h_{kl} to $h_{kl}/(1 - h_{jl})$ for $k \neq j$;
(iv) that eliminating risk j has no effect on the remaining hazards.

Assumptions (i) and (ii) together imply proportional hazards. In fact, we have then $h(j, t) = \omega_{jl}h(t)$ on I_l and so $f(j, t) = \omega_{jl}f(t)$, $\bar{F}(j, t) = \omega_{jl}\bar{F}(t)$, and $\omega_{+l} = \omega_{1l} + \ldots + \omega_{pl} = 1$.

Example 6.8 (David and Moeschberger, 1978, Section 5.3)

(a) Let

$$P_{(j)l} = \text{pr(survive } I_l \text{ under risk } j \text{ acting alone} \mid \text{enter } I_l)$$

$$= \text{pr}(T_j \geq \tau_l | T_j \geq \tau_{l-1}) = \bar{G}_j(\tau_l -)/\bar{G}_j(\tau_{l-1}-) = \exp\left\{-\int_{\tau_{l-1}}^{\tau_l} h_j(t)dt\right\}.$$

Then, under Chiang's proportionality assumption,

$$P_{(j)l} = \exp\left\{-\omega_{jl}\int_{\tau_{l-1}}^{\tau_l} h(t)dt\right\} = \{\bar{F}(\tau_l)/\bar{F}(\tau_{l-1})\}^{\omega_{jl}} = (1 - h_l)^{\omega_{jl}}.$$

We need an estimator for ω_{jl} since we already have $\hat{h}_l$. Under Assumptions (i) and (ii),

$$h_{jl} = \int_{\tau_{l-1}}^{\tau_l} f(j, t)dt/\bar{F}(\tau_{l-1}) = \int_{\tau_{l-1}}^{\tau_l} h(j, t)\bar{F}(t)dt/\bar{F}(\tau_{l-1})$$

$$= \omega_{jl}\int_{\tau_{l-1}}^{\tau_l} f(t)dt/\bar{F}(\tau_{l-1}) = \omega_{jl}h_l.$$

Thus, we have the estimate

$$\hat{\omega}_{jl} = \hat{h}_{jl}/\hat{h}_l = r_{jl}/r_l,$$

from which follows

$$\hat{P}_{(j)l} = (1 - r_{jl}/q_l)^{r_{jl}/r_l}.$$

Under a natural extension of Assumption (iii), h_{jl} would be replaced, in the absence of all other risks, by

$$h_{jl}/(1-\sum_{k\neq j} h_{kl}) = h_{jl}/(1-h_l+h_{jl}),$$

which can be estimated by $r_{jl}/(q_l - r_l + r_{jl})$. This yields, for $P_{(j)l}$, the estimate

$$\tilde{P}_{(j)l} = (q_l - r_l)/(q_l - r_l + r_{jl}) = \{1 + r_{jl}/(q_l - r_l)\}^{-1}.$$

(b) Let

$$P_{(-j)l} = \text{pr(survive } I_l \text{ with risk } j \text{ eliminated} \mid \text{enter } I_l).$$

Under Assumptions (i), (ii), and (iv),

$$P_{(-j)l} = \exp\left[-\int_{\tau_{l-1}}^{\tau_l} \{h(s) - h(j,s)\}ds\right] = \exp\left\{-(1-\omega_{jl})\int_{\tau_{l-1}}^{\tau_l} h(s)ds\right\}$$
$$= (1-h_l)^{1-\omega_{jl}}.$$

Hence,

$$\hat{P}_{(-j)l} = (1 - r_l/q_l)^{1 - r_{jl}/r_l}.$$

Under Assumption (iii),

$$P_{(-j)l} = 1 - \sum_{k\neq j} h_{kl}/(1-h_{jl}) = (1-h_l)/(1-h_{jl}),$$

which can be estimated by $(1 - r_l/q_l)/(1 - r_{jl}/q_l)$. ■

6.4 Proportional hazards and partial likelihood

This section gives a development of the methodology outlined in Section 4.3 to the Competing Risks context.

6.4.1 The proportional hazards model

By analogy with the proportional hazards model Equation 4.3.1 for univariate failure times, the sub-hazard functions for the ith case are specified as

$$h(j, t; x_i) = \psi_{ji} h_0(j, t), \tag{6.4.1}$$

where the $h_0(j, t)$ ($j = 1, \ldots, p$) form a set of baseline sub-hazards, and $\psi_{ji} = \psi_j(\boldsymbol{x}_i;\beta)$ is some positive function of $\boldsymbol{x}_i$ and β, respectively, a vector of explanatory variables and the associated vector of regression coefficients. A very common choice is $\psi_{ji} = \exp(\boldsymbol{x}_i^{\mathrm{T}}\beta_j)$, with β partitioned into subvectors as $(\beta_1, \ldots, \beta_p)$. This type of model was mentioned in Section 1.3.1: the difference here is that the $h_0(j, t)$ are left unspecified. In practical applications one would normally seek to limit the number of parameters by testing for restrictions on the β_js, e.g., that $\beta_1 = \beta_2$, or that particular components of β_3 are zero.

6.4.2 *The partial likelihood*

The observational set-up is as described in Section 6.3.1. The observed failure times from all causes are $t_1 < t_2 < \ldots < t_m$, with $t_0 = 0$ and $t_{m+1} = \infty$; R_{jl} is the set of r_{jl} individuals who fail from cause j at time t_l; S_l is the set of s_l individuals whose failure times t_{ls} ($s = 1, \ldots, s_l$) are right-censored during $[t_l, t_{l+1})$; $R(t_l)$ is the risk set (Section 4.1) at time t_l.

The probability that individual $i \in R(t_l)$ fails from cause j in the time interval $(t_l, t_l + dt]$ is $h(j, t_l;\boldsymbol{x}_i)dt$. Let $r_l = r_{1l} + \ldots + r_{pl}$ and suppose for the moment that $r_l = 1$, i.e., that there are no ties at time t_l. Suppose that t_l is the failure time of individual i_l, the cause being c_l. Given the events up to time t_l-, and given that there is a failure of type c_l at time t_l, the conditional probability that, among the members of $R(t_l)$, it is individual i_l at whom the famous fickle finger of fate and fortune points is

$$h(c_l, t_l;\boldsymbol{x}_{i_l})dt \div \sum\nolimits_l h(c_l, t_l;\boldsymbol{x}_a)dt = \psi_{c_l i_l} / \sum\nolimits_l \psi_{c_l a},$$

where $\sum_l$ denotes summation over individuals $a \in R(t_l)$. The corresponding partial likelihood function is, with d_l denoting $\sum_l \psi_{c_l a}$,

$$P(\beta) = \Pi_{l=1}^{m}(\psi_{c_l i_l}/d_l) \tag{6.4.2}$$

The maximum partial likelihood estimator $\hat{\beta}$ is found by maximizing $P(\beta)$ over β. Large-sample inference can be conducted by treating $\log P(\beta)$ as a log-likelihood function in the usual way; in the notation of Section 4.3.5 we take $A_l = \{i_l\}$ and $B_l = \{C_{l-1}, t_l, c_l\}$. Thus, we have

$$\log P(\beta) = \sum\nolimits_{l=1}^{m} \{\log \psi_{c_l i_l} - \log d_l\}$$

with first and second derivatives, respectively,

$$\begin{aligned} \boldsymbol{U}(\beta) &= \partial \log P(\beta)/\partial\beta = \sum\nolimits_{l=1}^{m} \boldsymbol{U}_l(\beta), \\ \boldsymbol{V}(\beta) &= -\partial^2 \log P(\beta)/\partial\beta^2 = \sum\nolimits_{l=1}^{m} \boldsymbol{V}_l(\beta), \end{aligned} \tag{6.4.3}$$

where the vector $U_l(\beta)$ has kth component $\partial\log(\psi_{c_l i_l}/d_l)/\partial\beta_k$ and the matrix $V_l(\beta)$ has kkth entry $-\partial^2\log(\psi_{c_l i_l}/d_l)/\partial\beta_k\partial\beta_{k'}$. With the usual choice, $\psi_{ji} = \exp(x_i^T\beta_j)$, we have $U_l(\beta) = x_{i_l} - v_l$ and $V_l(\beta) = X_l - v_l v_l^T$, where

$$v_l = d_l^{-1}\sum_l \psi_{c_l a} x_a \quad \text{and} \quad X_l = d_l^{-1}\sum_l \psi_{c_l a} x_a x_a^T.$$

Under some standard regularity conditions, $V(\hat{\beta})^{-1}$ provides an estimate for the variance matrix of $\hat{\beta}$, and so follow hypothesis tests and confidence intervals for β.

When there are tied values among the members of $R(t_l)$ the expression for $P(\beta)$ is more complicated. Thus, if there are $r_{jl} > 1$ observed failures from cause j at time t_l ($j = 1, \ldots, p$), the lth term in the product defining $P(\beta)$ should allow for all such distinct subsets of size r_{jl} selected from the individuals in the risk set $R(t_l)$. This can lead to an unwieldy computation, and a useful approximation is to replace this term by $\Pi(\psi_{ji_l}/d_l)$, where the product is taken over the r_{+l} tied cases. The quality of this approximation deteriorates as the proportion of tied values increases. Alternative options, like those listed in Section 4.3.2, are worth considering in this case.

Kay (1986) analyzed some data comprising 506 patients with prostate cancer each randomly allocated to 1 of 4 treatment regimes. There were seven additional explanatory variables including age and weight-height groups and disease history and severity indicators. There were three causes of death (cancer, cardiovascular, other), and time was recorded in months.

Goetghebeur and Ryan (1990, 1995) extended methods based on partial likelihood to cover the situation where some failure causes are missing in the data. In this case the numerator, $h(c_l, t_l; x_{i_l})dt$, in the equation preceding Equation 6.4.2 must be replaced by a sum of hazards over the possible failure causes. To maintain the vital cancellation of baseline hazards in this ratio, Goetghebeur and Ryan assumed that the baseline hazards for different causes are all proportional to a single baseline hazard function. This second proportionality assumption enables the development to go through as in the case of fully observed failure causes.

Example 6.9 Lagakos (1978)

We revisit the data of Lagakos (1978), previously fitted with a Weibull sub-hazards model in Example 2.6. Note that in $\psi_{ji} = \exp(x_i^T\beta_j)$ we do not need an intercept term here because the corresponding factor, $\exp(\beta_{0j})$, is absorbed into $h_0(t)$. The full model fit, applying maximum partial likelihood, yields estimates (0.65, 0.38, –2.07) for β_1 and (0.59, 0.44, 0.79) for β_2; the corresponding standard errors are (0.24, 0.28, 1.14) and (0.32, 0.39, 1.64). The previous estimates given in Example 2.6 were (0.47, 0.32, –1.26) for β_1 and (0.43, 0.32, 0.96) for β_2, with standard errors (0.17, 0.20, 0.84) and (0.23, 0.28, 1.18). The comparison could be said to reveal a certain qualitative resemblance. We do not have available likelihood ratio tests now, so we employ Wald tests. The first is to assess the null hypothesis that β_{21}, β_{31}, β_{22}, and β_{32} are all zero: this gives $\chi_4^2 = 6.05$ ($p > 0.10$), agreeing with the result in Example 2.6. However, a further Wald test for the hypothesis that all six βs are zero gives $\chi_6^2 = 6.33$ ($p > 0.10$), thus throwing doubt on the presence of an effect of any of the three explanatory variables. Incidentally, the equivalent Wald test based on the model fitted in Example 2.6 gives $\chi_6^2 = 2.56$ ($p > 0.10$), whereas a likelihood ratio test gives $\chi_6^2 = 14.33$ ($0.025 < p < 0.05$). It seems that the theoretical asymptotic equivalence of the Wald and likelihood ratio tests is not well demonstrated by these data coupled with the Weibull hazards model, though the Wald test is fairly consistent between the Weibull hazards model and the semi-parametric proportional hazards model. One might say that the picture is not crystal clear. ■

6.4.3 The baseline survivor functions

The likelihood function, like Equation 6.3.1 but with explanatory variables, is

$$L = \Pi_{l=1}^{m}\Pi_{i \in R_l}\Delta\bar{F}(c_l, t_l;\boldsymbol{x}_i) \times \Pi_{l=0}^{m}\Pi_{i \in S_l}\bar{F}(t_{ls};\boldsymbol{x}_i) \quad . \tag{6.4.4}$$

To maximize L we must take $\hat{\bar{F}}(t_{ls};\boldsymbol{x}_i)$ equal to $\hat{\bar{F}}(t_l;\boldsymbol{x}_i)$ for $i \in S_l$ (as explained in Section 4.2.1), for which we must take $\hat{\bar{F}}(j, t_{ls};\boldsymbol{x}_i) = \hat{\bar{F}}(j, t_l;\boldsymbol{x}_i)$ for each j. In particular, $\hat{\bar{F}}(t_{0s}) = 1$ which, in effect, removes the factor for $l = 0$ from the second term in Equation 6.4.4. Also, $\hat{\bar{F}}(c_i, t;\boldsymbol{x}_i)$ must have a discontinuity at $t = t_l$ for $i \in R_l$, otherwise $L = 0$. Therefore, $\hat{\bar{F}}(t;\boldsymbol{x}_i)$, which equals $\sum_{j=1}^{p}\hat{\bar{F}}(j, t;\boldsymbol{x}_i)$, also has such a discontinuity. Take

$$\hat{\bar{F}}(t_l;\boldsymbol{x}_i) = \Pi_{s=1}^{l}\{1 - \hat{h}(t_s;\boldsymbol{x}_i)\},$$

which allows for discontinuities at each t_s, and use

$$h(t_s;\boldsymbol{x}_i) = \sum_{j=1}^{p} h(j, t_s;\boldsymbol{x}_i) = \sum_{j=1}^{p}\psi_{ji}h_0(j, t_s),$$

which follows from Equation 6.4.1. Then

$$\hat{\bar{F}}(t_l;\boldsymbol{x}_i) = \prod_{s=1}^{l}\left\{1 - \sum_{j=1}^{p}\psi_{ji}\hat{h}_0(j, t_s)\right\},$$

and

$$\Delta\hat{\bar{F}}(j, t_l;\boldsymbol{x}_i) = \hat{h}(j, t_l;\boldsymbol{x}_i)\hat{\bar{F}}(t_l-;\boldsymbol{x}_i) = \psi_{ji}\hat{h}_0(j, t_l)\prod_{s=1}^{l-1}\left\{1 - \sum_{j=1}^{p}\psi_{ji}\hat{h}_0(j, t_s)\right\}.$$

Hence, as in Equation 4.1.7 and with Q_l defined in Equation 4.1.8, we have to maximise

$$L = \prod_{l=1}^{m} \prod_{i \in R_l} \left[\psi_{c_i i} h_0(c_i, t_l) \Pi_{s=1}^{l-1} \left\{ 1 - \sum_{j=1}^{p} \psi_{ji} h_0(j, t_s) \right\} \right]$$

$$\times \prod_{l=1}^{m} \prod_{i \in S_l} \prod_{s=1}^{l} \left\{ 1 - \sum_{j=1}^{p} \psi_{ji} h_0(j, t_s) \right\}$$

$$= \prod_{l=1}^{m} \left[\prod_{i \in R_l} \psi_{c_i i} h_0(c_i, t_l) \times \prod_{i \in Q_l} \left\{ 1 - \prod_{j=1}^{p} \psi_{ji} h_0(j, t_l) \right\} \right]$$

over the "parameters" $h_0(j, t_s)$ ($j = 1, \ldots, p$; $s = 1, \ldots, m$), having replaced ψ_{ji} by $\psi_j(\boldsymbol{x}_i;\hat{\beta})$ from the partial likelihood estimation.

The function is orthogonal in the parameter sets $\boldsymbol{h}_l = \{h_0(j, t_l){:}j = 1, \ldots, p\}$, so the numerical maximisation is simplified (Appendix 2). Armed with the estimates $\hat{h}_0(j, t_s)$, those for the baseline survivor functions follow from

$$\hat{\bar{F}}_0(t_l) = \Pi_{s=1}^{l}\{1 - \hat{h}_0(t_s)\}, \ \Delta\hat{\bar{F}}_0(j, t_l) = \hat{h}_0(j, t_l)\hat{\bar{F}}_0(t_{l-1}),$$

where $\hat{h}_0(t_s) = \sum_{j=1}^{p} h_0(j, t_s)$.

6.4.4 *Discrete failure times*

We will just describe the extension of Section 4.4.2 to Competing Risks. The argument for constructing the partial likelihood is the same as in Section 6.4.2. Suppose for the moment that individual i_l is the only one to fail at time τ_l, and that the cause of failure is c_l. Then the lth contribution to the partial likelihood function is

$$[h_l(c_l;\boldsymbol{x}_{i_l})/\{1 - h_l(c_l;\boldsymbol{x}_{i_l})\}] \div \sum_{i \in R(\tau_l)} [h_l(c_l;\boldsymbol{x}_i)/\{1 - h_l(c_l;\boldsymbol{x}_i)\}]$$

as in Equation 4.4.3, writing $h_l(j;\boldsymbol{x})$ for $h(j, \tau_l;\boldsymbol{x})$. The assumption of a proportional odds model

$$[h_l(j;\boldsymbol{x}_i)/\{1 - h_l(j;\boldsymbol{x}_i)\}] = \psi_{ji}[h_0(j, l)/\{1 - h_0(j, l)\}],$$

where $\psi_{ji} = \psi_j(\boldsymbol{x}_i;\beta)$, reduces this contribution to $\psi_{c_l i_l}/d_l$, d_l denoting $\sum_{i \in R(\tau_l)} \psi_{c_l i_l}$. Hence, the partial likelihood is

$$P(\beta) = \Pi_{l=1}^{m}(\psi_{c_l i_l}/d_l)$$

and it is put to use as in Section 6.4.2.

Ties are a nuisance and must be dealt with along the lines set out in Section 4.4.2.

For estimation of the baseline hazards $h_0(j, l)$ the relevant likelihood function is derived in Section 5.5. Thus, we take L as in Equation 5.5.3 with $h(c_i, \tau_l; \boldsymbol{x}_i)$ there replaced by

$$\psi_{ji} h_0(c_i, l) / \{1 + (\psi_{ji} - 1) h_0(c_i, l)\},$$

according to the proportional odds model; the replacement for $h(\tau_l; \boldsymbol{x}_i)$ in Equation 5.5.3 follows from its expression as $\sum_j h(j, \tau_l; \boldsymbol{x}_i)$. In ψ_{ji}, β can be replaced by $\hat{\beta}$, the maximum partial likelihood estimator, and then L is to be maximised over the remaining parameter set $\{h_0(j, l): j = 1, \ldots, p; l = 1, \ldots, m\}$.

chapter 7

Latent failure times: identifiability crises

Looking at Competing Risks from the point of view of latent lifetimes seems very natural. Statistically respectably, you can specify a parametric model for $\bar{G}(t)$, fit it to the data, test the fit and test hypotheses, all without the faintest suspicion that there might be anything lacking. A hidden problem with this approach is explored in this chapter.

7.1 The Cox-Tsiatis impasse

Cox (1959) drew attention to a flaw in the approach to Competing Risks via latent lifetime models. His discussion concerned various models for failure times with two causes, like those presented by Mendenhall and Hader (1958), including their mixture of exponentials in Example 1.2. He noted that "no data of the present type can be inconsistent with" an independent-risks model. This was extended to the general case of p risks by Tsiatis (1975). Specifically, Tsiatis showed that, given any joint survivor function with arbitrary dependence between the component variates, there exists a different joint survivor function in which the variates are independent and which reproduces the sub-densities $f(j, t)$ precisely. Thus, one cannot know, from observations on (C, T) alone, which of the two models is correct — they will both fit the data equally well. This is a bit awkward from the point of view of statistical inference.

Theorem 7.1 (Tsiatis, 1975)

Suppose that the set of $\bar{F}(j, t)$ is given for some model with dependent risks. Then there exists a unique **proxy model** with independent risks yielding identical $\bar{F}(j, t)$. It is defined by $\bar{G}^*(t) = \Pi_{j=1}^{p} \bar{G}_j^*(t_j)$, where $\bar{G}_j^*(t) = \exp\{-\int_0^t h(j, s)ds\}$ and the sub-hazard function $h(j, s)$ derives from the given $\bar{F}(j, t)$.

Proof

A constructive proof will be given. We wish to find a function $\bar{G}^*(\mathbf{t}) = \Pi_{j=1}^p \bar{G}_j^*(t_j)$ such that $f(j, t) = [-\partial \bar{G}^*(\mathbf{t})/\partial t_j]_{t\mathbf{1}_p}$ for all (j, t). Hence, we require that

$$f(j,t) = [\{-\partial \bar{G}_j^*(t_j)/\partial t_j\}\bar{G}^*(\mathbf{t})/\bar{G}_j^*(t_j)]_{t\mathbf{1}_p} \\ = \{-d\log \bar{G}_j^*(t)/dt\}\bar{G}^*(t\mathbf{1}_p). \tag{7.1.1}$$

Summing Equation 7.1.1 over j yields

$$f(t) = \{-d\log \bar{G}^*(t\mathbf{1}_p)/dt\}\bar{G}^*(t\mathbf{1}_p) = -d\bar{G}^*(t\mathbf{1}_p)/dt.$$

Integrating, $\bar{F}(t) = \bar{G}^*(t\mathbf{1}_p)$ and so, from Equation 7.1.1,

$$-d\log \bar{G}_j^*(t)/dt = f(j,t)/\bar{F}(t) = h(j,t).$$

We have shown so far that, if an independent-risks proxy model exists, then it must have the given form. To complete the proof, we have to show that the given form defines $\bar{G}_j^*(t)$ as a valid survivor function, which it obviously does, and also meets the requirement of mimicking the $f(j, t)$. For this last part

$$g^*(j,t) = g_j^*(t)\Pi_{k\neq j}\bar{G}_k^*(t) = \{-d\log \bar{G}_j^*(t)/dt\}\Pi_{k=1}^p\bar{G}_k^*(t) \\ = h(j,t)\exp\left\{-\Pi_{k=1}^p\int_0^t h(k,s)ds\right\} = h(j,t)\exp\left\{-\int_0^t h(s)ds\right\} \\ = h(j,t)\bar{F}(t) = f(j,t). \quad \blacksquare$$

From Equation 1.1.1, the hazard function h_j^* corresponding to $\bar{G}_j^*$ satisfies

$$\bar{G}_j^*(t) = \exp\left\{-\int_0^t h_j^*(s)ds\right\}.$$

Thus, the proxy model of the theorem is just constructed by using the original sub-hazards $h(j, t)$ for the $h_j^*(t)$. If the original model has independent risks then it is its own proxy because in this case $h(j, t) = h_j(t)$, by Theorem 3.3, and so $\bar{G}_j^*(t) = \bar{G}_j(t)$. This hazard condition can hold even without independent risks, as noted in Section 3.4.1, so some dependent-risks systems can provide their own proxy models. This curiosity will be taken up more formally in Section 7.3. Under the classical assumption of independent risks what one is actually estimating is the proxy model, because one observes

$h(j, t)$ and assumes that this is $h_j(t)$. Thus, there's an outside chance of being correct, but not one that you'd want to bet on.

In the case of proportional hazards,

$$\bar{G}_j^*(t) = \exp\left\{-p_j\int_0^t h(s)ds\right\} = \exp\{p_j \log \bar{F}(t)\} = \bar{F}(t)^{p_j},$$

yielding an expression for $\bar{G}^*(\boldsymbol{t})$ in terms of the single function $\bar{F}(t)$; this was noted previously in Lemma 3.4.

Theorem 7.1 establishes only that to each dependent-risks model there corresponds a unique independent-risks proxy model with the same sub-survivor functions. More structure will be revealed in Section 7.3 where it will be shown that each independent-risks model has a whole class of satellite dependent-risks models and that this class can be further partitioned into sets with the same marginals. The following examples illustrate this structure within the limited framework of particular parametric classes of models.

Example 7.1 Bivariate exponential (Gumbel, 1960; Example 3.1)

$$\bar{G}(\boldsymbol{t}) = \exp\{-\lambda_1 t_1 - \lambda_2 t_2 - \nu t_1 t_2\}.$$

The sub-hazard functions are $h(j, t) = \lambda_j + \nu t$ (Section 3.1) and so the proxy model has

$$\bar{G}_j^*(t) = \exp\left\{-\int_0^t (\lambda_j + \nu s)ds\right\} = \exp\{-(\lambda_j t + \nu t^2/2)\},$$

$$\bar{G}^*(\boldsymbol{t}) = \exp\{-(\lambda_1 t_1 + \nu t_1^2/2) - (\lambda_2 t_2 + \nu t_2^2/2)\}.$$

Hence, predictions about T_j based on $\bar{G}_j(t) = \exp(-\lambda_j t)$ would differ from those based on $\bar{G}_j^*(t)$, and you cannot tell which function is the correct one to use purely from (c, t)-data, however much of it you have. ■

There is a 1-1 correspondence between $\bar{G}(\boldsymbol{t})$ and $\bar{G}^*(\boldsymbol{t})$ in the example because the parameter set $(\lambda_1, \lambda_2, \nu)$ is identified by both $\bar{G}(\boldsymbol{t})$ and $\bar{G}^*(\boldsymbol{t})$. But it ain't necessarily so, as shown by a second example.

Example 7.2 (Crowder, 1993)

Consider the form

$$\bar{G}(\boldsymbol{t}) = \exp\{-\lambda_1 t_1 - \lambda_2 t_2 - \nu t_1 t_2 - \mu_1 t_1^2 - \mu_2 t_2^2\}.$$

This is a valid bivariate survivor function if $\lambda_j > 0$ and $\mu_j \geq 0$ for $j = 1, 2$, and $0 \leq \nu \leq \lambda_1\lambda_2$. The sub-density and sub-hazard functions are

$$f(j, t) = \{\lambda_j + (\nu + 2\mu_j)t\}\exp\{-\lambda_+ t - (\nu + \mu_+)t^2\},\ h(j, t) = \lambda_j + (\nu + 2\mu_j)t,$$

where $\lambda_+ = \lambda_1 + \lambda_2$ and $\mu_+ = \mu_1 + \mu_2$. The proxy model then has marginals

$$\bar{G}_j^*(t) = \exp\left\{-\int_0^t h(j, s)ds\right\} = \exp\{-\lambda_j t - (\nu + 2\mu_j)t^2/2\}.$$

Evidently, $\bar{G}^*(\boldsymbol{t})$ identifies only $(\lambda_1, \lambda_2, \nu + 2\mu_1, \nu + 2\mu_2)$, so the set of $\bar{G}(\boldsymbol{t})$ with $\nu + 2\mu_1 = \tau_1$ and $\nu + 2\mu_2 = \tau_2$ all share the same independent-risks proxy model

$$\bar{G}^*(\boldsymbol{t}) = \exp\{-\lambda_1 t_1 - \lambda_2 t_2 - \tau_1 t_1^2/2 - \tau_2 t_2^2/2\}. \blacksquare$$

The following third example gives a general class of parametric models in which the non-identifiability behaviour occurs.

Example 7.3 (Crowder, 1991)

Consider a joint density of the form

$$g(\boldsymbol{t}) = \begin{cases} g_\phi(t_2)g_\psi(t_1 - t_2)/2 & \text{on } t_1 > t_2, \\ g_\phi(t_1)g_\psi(t_2 - t_1)/2 & \text{on } t_1 < t_2, \end{cases}$$

where g_ϕ and g_ψ are univariate densities on $(0, \infty)$ involving parameters ϕ and ψ. By the formula $f(j, t) = \int_t^\infty g(\boldsymbol{r}_j)ds$ (Section 3.1), we have $f(1, t) = g_\phi(t)/2$ and, symmetrically, $f(2, t) = g_\phi(t)/2$. Thus, ψ has gone absent without leave from the sub-distributions. A particular case is obtained when g_ϕ and g_ψ are exponential densities. In that case,

$$\bar{G}(\boldsymbol{t}) = \begin{cases} (1+\tau)\exp(-\phi t_1) - \tau\exp(-\psi t_1 + \psi t_2 - \phi t_2) & \text{on } t_1 > t_2, \\ (1+\tau)\exp(-\phi t_2) - \tau\exp(-\psi t_2 + \psi t_1 - \phi t_1) & \text{on } t_1 < t_2, \end{cases}$$

where $\tau = \phi/\{2(\psi - \phi)\}$. It can be verified that $\theta = (\phi, \psi)$ is identified in the joint distribution but that $\bar{F}(1, t) = \bar{F}(2, t) = e^{-\phi t}$, independent of ψ. $\blacksquare$

7.2 More general identifiability results

The assumption that T_j is continuous in Theorem 7.1 can be dropped. As before, $T = \min\{T_1, \ldots, T_p\}$, and let $\alpha = \sup\{t: \bar{F}(t) > 0\}$.

Theorem 7.2 (Miller, 1977)

Suppose that the $F(j, t)$ $(j = 1, \ldots, p)$ have no discontinuities in common and there are no ties amongst the T_j. Then there exists a set of independent $S_j (j = 1, \ldots, p)$, at least one of which is almost surely finite, such that their sub-survivor functions match the $\bar{F}(j, t)$: $\bar{G}_s(j, t) = \bar{F}(j, t)$ for all (j, t). Also, the S_j- distributions are uniquely determined on $[0, \alpha)$.

Proof

See the paper cited. ■

The assumption that there are no ties can also be dropped. Typical applications occur in Reliability where a piece of equipment is subject to random shocks which can knock out one or more components at the same time. In this extended situation, we still have $T = \min\{T_1, \ldots, T_p\}$ but C is defined more generally as the **failure pattern** or **configuration** (Section 5.2). A simple but significant point is that ties between configurations are eliminated by definition: for example, instead of saying that $\{j, k\}$ and $\{l\}$ have tied, we simply say that $\{j, k, l\}$ has occurred.

Let $\boldsymbol{R}$ be a failure time vector of length $2^p - 1$ with components R_C, and define C_R and T_R by $T_R = \min_c(R_c) = R_{C_R}$. The next theorem concerns the existence and form of an independent-risks proxy model such that

$$(C_R, T_R) \overset{d}{=} (C, T),$$

i.e., the pair (C_R, T_R) from the $\boldsymbol{R}$-system has the same joint distribution as the original (C, T). In the authors' alternative terminology, systems $\boldsymbol{R}$ and $\boldsymbol{T}$ are *equivalent in life length and patterns*, written as $\boldsymbol{R} = {}_{LP}\boldsymbol{T}$, if $\text{pr}(C_R = c, T_R > t) = \bar{F}(c, t)$ for all configurations c and all $t \geq 0$. The result extends Miller's (1977) by allowing ties and also giving explicit forms for the R_c-distributions. Peterson (1977) covered similar ground, but with a different emphasis.

Before stating the theorem, we need some notation. Let $D_c = \{\tau_{cl}: l = 1, 2, \ldots\}$ be the set of discontinuities of $\bar{F}(c, t)$, let the discrete sub-hazard contribution at τ_{cl} be

$$h_{cl} = \text{pr}(C = c, T = \tau_{cl} \mid T > \tau_{cl}-) = \Delta\bar{F}(c, \tau_{cl}) / \bar{F}(\tau_{cl}-),$$

where

$$\Delta\bar{F}(c, \tau_{cl}) = \bar{F}(c, \tau_{cl}-) - \bar{F}(c, \tau_{cl}),$$

and let, for $t \notin D_c$,

$$f^c(c, t) = -d\bar{F}^c(c, t)/dt \quad \text{and} \quad h^c(c, t) = f^c(c, t)/\bar{F}(t-)$$

be the density and hazard functions of the continuous component of $\bar{F}(c, t)$.

Theorem 7.3 (Langberg, Proschan and Quinzi, 1978)

There exists a set of independent R_c with $\boldsymbol{R} = {}_{LP}\boldsymbol{T}$ if and only if the D_c are pairwise disjoint on $[0, \alpha)$. In this case, the survivor functions of the R_c are uniquely determined on $[0, \alpha)$ as

$$\bar{G}_{Rc}(t) = \text{pr}(R_c > t) = \exp\left\{-\int_0^t h^c(c, s)ds\right\} \Pi_{s=1}^{l_c(t)}(1 - h_{cs}), \qquad (7.2.1)$$

where $l_c(t) = \max\{l{:}\tau_{cl} \leq t\}$.

Proof

That disjoint D_c are necessary can be seen as follows. Suppose that they are not disjoint, in particular that $\tau \in D_b \cap D_c$. If an independent-risks $\boldsymbol{R}$-system existed which matched the original sub-survivor functions, we would have $\bar{F}_R(c, t) = \bar{F}(c, t)$ for all (c, t), and so $\bar{F}_R(b, t)$ and $\bar{F}_R(c, t)$ would both be discontinuous at τ. In that case $\text{pr}(R_b = \tau) > 0$ and $\text{pr}(R_c = \tau) > 0$, and so $\text{pr}(E_{bc}) > 0$, where $E_{bc} = \{R_b = R_c = \tau < \text{all other } Rs\}$, since the Rs are independent. Consequently, $\sum_c \bar{F}_R(c, 0) < 1$ because the sum does not include $\text{pr}(E_{bc})$. Conversely, under matching of the sub-survivor functions,

$$\sum_c \bar{F}_R(c, 0) = \sum_c \bar{F}(c, 0) = 1.$$

Now assume that the D_c are disjoint. The form of Equation 7.2.1 can be guessed by noting that the proxy model in Theorem 7.1 is obtained by using the original sub-hazards for the proxy marginal hazards, and then extending this to cover discontinuities via the standard formula Equation 4.1.3. We just have to prove now that it does the job as advertised, i.e., that $\bar{F}_R(c, t) = \bar{F}(c, t)$ for all (c, t). Now,

$$\bar{F}_R(c, t) = \int_t^\infty g_{Rc}^c(s)\left\{\prod_{b \neq c} \bar{G}_{Rb}(s)\right\}ds + \sum_{s > l(t)} \{\Delta\bar{G}_{Rc}(\tau_{cs})\Pi_{b \neq c}\bar{G}_{rb}(\tau_{cs})\}, \quad (7.2.2)$$

where $\Delta\bar{G}_{Rc}(\tau_{cs}) = \bar{G}_{Rc}(\tau_{cs}-) - \bar{G}_{Rc}(\tau_{cs})$, and $g_{Rc}^c(s) = -d\bar{G}_{Rc}(s)/ds$ is the marginal density defined for $s \notin D_c$. First, from Equation 7.2.1,

$$\Pi_b\bar{G}_{Rb}(t) = \exp\left\{-\int_0^t h^c(s)ds\right\} \times \Pi_b\Pi_{s=1}^{l_b(t)}(1 - h_{bs}) = \bar{F}(t), \qquad (7.2.3)$$

using $h^c(s) = \sum_b h^c(b, s)$ and the fact that the product is over all the discontinuities of $\bar{F}(t) = \sum_b \bar{F}(b, t)$ up to t. The first term on the right hand side of Equation 7.2.2 is then equal to

$$\int_t^{\infty} h_{Rc}^c(s)\{\Pi_b \bar{G}_{Rb}(s)\}ds = \int_t^{\infty} h^c(c, s)\bar{F}(s)ds = \int_t^{\infty} f^c(c, s)ds = \bar{F}^c(c, t),$$

where we have used

$$h_{Rc}^c(s) = -d\log \bar{G}_{Rc}^c(s)/ds = h^c(c, s),$$

from Equation 7.2.1. For the second term in Equation 7.2.2 we have, from Equation 7.2.1,

$$\Delta\bar{G}_{Rc}(\tau_{cl}) = \exp\left\{-\int_0^{\tau_{cl}} h^c(c, s)ds\right\} \times \{\Pi_{s=1}^{l-1}(1-h_{cs}) - \Pi_{s=1}^{l}(1-h_{cs})\}$$

$$= \exp\left\{-\int_0^{\tau_{cl}} h^c(s)ds\right\} \times \{h_{cl}/(1-h_{cl})\}\Pi_{s=1}^{l}(1-h_{cs})$$

$$= \bar{G}_{Rc}(\tau_{cl})\{h_{cl}/(1-h_{cl})\}.$$

Further,

$$h_{cl}/(1-h_{cl}) = \{\Delta\bar{F}(c, \tau_{cl})/\bar{F}(\tau_{cl}-)\} \div \{1 - \Delta\bar{F}(c, \tau_{cl})/\bar{F}(\tau_{cl}-)\}$$

$$= \Delta\bar{F}(c, \tau_{cl}) \div \{\bar{F}(\tau_{cl}-) - \Delta\bar{F}(c, \tau_{cl})\}.$$

But,

$$\Delta\bar{F}(\tau_{cl}) = \Delta\sum_b \bar{F}(b, \tau_{cl}) = \Delta\bar{F}(c, \tau_{cl}),$$

since $\Delta\bar{F}(b, \tau_{cl}) = 0$, in consequence of $\tau_{cl} \notin D_b$ when $b \neq c$. So,

$$h_{cl}/(1-h_{cl}) = \Delta\bar{F}(c, \tau_{cl}) \div \{\bar{F}(\tau_{cl}-) - \Delta\bar{F}(\tau_{cl})\} = \Delta\bar{F}(c, \tau_{cl}) \div \bar{F}(\tau_{cl}).$$

Thus, the second term in Equation 7.2.2 is equal to

$$\sum_{s>l(t)} [\bar{G}_{Rc}(\tau_{cs})\{\Delta\bar{F}(c, \tau_{cs})/\bar{F}(\tau_{cs})\}\prod_{b\neq c}\bar{G}_{Rb}(\tau_{cs})] = \Pi_{s>l(t)}\Delta\bar{F}(c, \tau_{cs}),$$

using Equation 7.2.3. Putting the first and second terms together, Equation 7.2.2 is $\bar{F}(c, t)$, as advertised. ■

Arjas and Greenwood (1981) extended Theorem 7.3. They allowed the set of configurations to be countable, rather than just finite, and showed that the disjoint D_c condition can be avoided by using a random tie-breaking device. Their presentation is in terms of martingales and compensators, but a rough translation is as follows. First, let $\bar{G}_{Rc}(t)$ be given by Equation 7.2.1, and define

$\hat{\bar{G}}_{Rc}(t) = \bar{G}_{Rc}(t)$ except when t is a discontinuity point common to more than one D_c. At such a point, say τ, let $d_\tau = \sum_c \Delta\bar{G}_{Rc}(\tau)$ be the total jump and reallocate the whole of d_τ to just one of the $\hat{\bar{G}}_{Rc}$ at τ. Theorem 7.3 now applies to this new system because the modified discontinuity sets are disjoint. Thus, we have a set of independent R_c whose sub-survivor functions match the modified $\bar{F}(c, t)$. Cunningly, Arjas and Greenwood make the modification randomly. They reallocate $\hat{d}_\tau$ with probabilities $q_{c\tau} = \Delta\bar{G}_{Rc}(\tau)/d_\tau$: $\Delta\hat{\bar{G}}_{Rc}(\tau)$ becomes 0 or d_τ with $\text{pr}\{\Delta\hat{\bar{G}}_{Rc}(\tau) = d_\tau\} = q_c\tau$. The R_c are then conditionally independent, given the particular reallocation, and, unconditionally,

$$\text{pr}(R_c > t) = \text{E}\{\hat{\bar{G}}_{Rc}(t)\} = \text{E}\left\{\hat{\bar{G}}^c_{Rc}(t) - \sum_{s=1}^{l(t)} \Delta\hat{\bar{G}}_{Rc}(\tau_{cs})\right\}$$

$$= \bar{G}^c_{Rc}(t) - \sum_{s=1}^{l(t)} \Delta\bar{G}_{Rc}(\tau_{cs}) = \bar{G}_{Rc}(t).$$

The R_c are unconditionally dependent: for instance, if $\tau \in D_b \cap D_c$ then $\text{pr}(S_c = \tau \mid S_b = t)$ depends on t by being 0 if $t = \tau$ but not otherwise.

Given an $\boldsymbol{R}$-system, a failure time vector $\boldsymbol{S} = (S_1, \ldots, S_p)$ can be constructed such that $\boldsymbol{R} = {}_{LP}\boldsymbol{S}$ as follows. Define $S_j = \min_c\{R_c : c \ni j\}$, so $C_S = c$ if and only if $C_R = c$, i.e., configuration c fails first, knocking out all components j with $j \in c$. For example, suppose that an $\boldsymbol{R}$-system comprises R_1, R_2, R_3, $R_{\{1,3\}}$ and $R_{\{1,2,3\}}$. Then we define $S_1 = \min\{R_1, R_{\{1,3\}}, R_{\{1,2,3\}}\}$, $S_2 = \min\{R_2, R_{\{1,2,3\}}\}$, and $S_3 = \min\{R_3, R_{\{1,3\}}, R_{\{1,2,3\}}\}$. Hence,

$$\bar{G}_S(c, t) = \text{pr}(C_S = c, T_S > t) = \text{pr}(C_R = c, T_R > t),$$

i.e., $\boldsymbol{R} = {}_{LP}\boldsymbol{S}$.

The $\boldsymbol{S}$-system has p components, like the original $\boldsymbol{T}$ but unlike the vaguely artificial $2^p - 1$ components of $\boldsymbol{R}$, and the S_j will be dependent in general. Their joint survivor function may be expressed in terms of the $\boldsymbol{R}$-system as follows:

$$\bar{G}_S(\boldsymbol{s}) = \text{pr}[\cap_{j=1}^p \{S_j > s_j\}] = \text{pr}[\cap_{j=1}^p \{\min_{c \ni j}(R_c) > s_j\}]$$
$$= \text{pr}[\cap_c \{R_c > \max_{j \in c}(s_j)\}] = \Pi_c \text{pr}\{R_c > \max_{j \in c}(s_j)\}.$$

The marginal survivor function of S_j is then found by setting all the s_k, other than s_j, to zero in $\bar{G}_S(\boldsymbol{s})$:

$$\bar{G}_{Sj}(s_j) = \Pi_{c \ni j} \text{pr}(R_c > s_j).$$

One can believe that the S_j are dependent in general by staring hard at their definition given above. However, a lurking doubt might remain as to whether they might, in some special circumstance that you can't quite put your finger on, achieve independence. The following lemma can bring express relief to this discomfort.

Lemma 7.4

The S_j are either dependent or degenerate.

Proof

The dependence coefficient defined in Section 3.1.1 is, for the **S**-system,

$$\bar{R}(\mathbf{s}) = \text{pr}(\mathbf{S} > \mathbf{s}) \div \Pi_{j=1}^{p} \text{pr}(S_j > s_j) = \Pi_c \text{pr}(R_c > m_c) \div \Pi_{j=1}^{p} \Pi_{c \ni j} \text{pr}(R_c > s_j),$$

where $m_c = \max_{j \in c}(s_j)$. The numerator here is equal to

$$\Pi_c \left[\{\Pi_{\tau_{cl} \le m_c}(1 - h_{cl})\} \exp\left\{ -\int_0^{m_c} h^c(c, s) ds \right\} \right],$$

and, using $\Pi_{j=1}^{p} \Pi_{c \ni j} = \Pi_c \Pi_{j \in c}$, the denominator is equal to

$$\Pi_c \Pi_{j \in c} \text{pr}(R_c > s_j) = \Pi_c \Pi_{j \in c} \left[\{\Pi_{\tau_{cl} \le s_j}(1 - h_{cl})\} \exp\left\{ -\int_0^{s_j} h^c(c, s) ds \right\} \right]$$

$$= \Pi_c \left[\{\Pi_l (1 - h_{cl})^{q_{ck}}\} \exp\left\{ -\int_0^{m_c} q_c(s) h^c(c, s) ds \right\} \right],$$

where $q_{cl} = \sum_{j \in c} \text{I}\{\tau_{cl} \le s_j\}$ and $q_c(s) = \sum_{j \in c} \text{I}\{s \le s_j\}$. Hence,

$$\bar{R}(\mathbf{s}) = \Pi_c \left[\{\Pi(1 - h_{cl})^{1 - q_{cl}}\} \exp\left\{ -\int_0^{m_c} \{1 - q_c(s)\} h^c(c, s) ds \right\} \right],$$

where the bracketed product is over $\{l: \tau_{cl} \le m_c\}$. Note that $q_{cl} \ge 1$ on $\{l: \tau_{cl} \le m_c\}$ and $q_c(s) \ge 1$ on $(0, m_c)$; it follows that $(1 - h_{cl})^{1 - q_{cl}} \ge 1$ and

$$\int_0^{m_c} \{1 - q_c(s)\} h^c(c, s) ds \le 0.$$

Hence, for $\bar{R}(\mathbf{s})$ to be 1, we must have both $(1 - h_{cl})^{1 - q_{cl}} = 1$ for all (c, l) and

$$\int_0^{m_c} \{1 - q_c(s)\} h^c(c, s) ds = 0$$

for all c, i.e., both (i) either $h_{cl} = 0$ or $q_{cl} = 1$, and (ii) either $h^c(c, s) = 0$ or $q_c(s) = 1$. In (i), by definition $h_{cl} > 0$ and, since we can choose s_j to make $q_{cl} \neq 1$, the product term must be absent, i.e., there can be no discontinuities. In (ii), since we can choose s_j to make $q_c(s) \neq 1$, we must have $h^c(c, s) = 0$ for all s. Thus, the only way that $\bar{R}(\boldsymbol{s})$ can be 1 for all $\boldsymbol{s}$ is for there to be no discontinuities and $h^c(c, s) = 0$ for all (c, s), in which case $\bar{F}(c, t) = 1$ for all (c, t), a sadly degenerate case. ■

Example 7.4 Bivariate exponential (Marshall and Olkin, 1967a)

Consider a two-component system subject to three types of fatal shock. The first type knocks out Component 1; the second, Component 2; and the third, both components. Suppose that the shocks arrive according to three independent Poisson processes of rates λ_1, λ_2 and λ_{12}, and let $N_j(t)$ be the number of shocks of type j occurring during time interval $(0, t)$. Then

$$\begin{aligned}\bar{G}(t) &= \text{pr}(T_1 > t_1, T_2 > t_2) = \text{pr}[N_1(t_1) = 0, N_2(t_2) = 0, N_3\{\max(t_1, t_2)\} = 0] \\ &= e^{-\lambda_1 t_1} e^{-\lambda_2 t_2} e^{-\lambda_{12} \max(t_1, t_2)}.\end{aligned}$$

An equivalent representation is in terms of the exponential waiting times W_j $(j = 1, 2, 3)$ to the three types of shock: $T_1 = \min\{W_1, W_3\}$ and $T_2 = \min\{W_2, W_3\}$. Marshall and Olkin gave two other derivations of this distribution, one via a different Poisson-shocks model and the other via the exponential lack-of-memory characterization.

The distribution has both a continuous and a singular component. The latter arises from the non-zero probability mass $\text{pr}(T_1 = T_2) = \lambda_{12}/\lambda_+$, where $\lambda_+ = \lambda_1 + \lambda_2 + \lambda_{12}$, concentrated on the subspace $\{\boldsymbol{t}: t_1 = t_2\}$ of $(0, \infty)^2$. The precise forms of these components can be calculated as follows. We first find the continuous component of the distribution, i.e., that which has a density. Now

$$\partial^2 \bar{G}(\boldsymbol{t})/\partial t_1 \partial t_2 = \begin{cases} \lambda_2(\lambda_1 + \lambda_{12}) \exp(-\lambda_1 t_1 - \lambda_2 t_2 - \lambda_{12} t_1) = g_1(\boldsymbol{t}) & \text{on } t_1 > t_2, \\ \lambda_1(\lambda_2 + \lambda_{12}) \exp(-\lambda_1 t_1 - \lambda_2 t_2 - \lambda_{12} t_1) = g_2(\boldsymbol{t}) & \text{on } t_1 < t_2. \end{cases}$$

Integrating,

$$\begin{aligned}\bar{G}^c(\boldsymbol{t}) &= \int_{t_1}^{\infty} dt_1 \int_{t_2}^{\infty} dt_2 \{\partial^2 \bar{G}(\boldsymbol{t})/\partial t_1 \partial t_2\} \\ &= \int_{t_1}^{\infty} dt_1 \int_{t_2}^{t_1} dt_2 \{g_1(\boldsymbol{t})\} + \int_{t_1}^{\infty} dt_1 \int_{t_1}^{\infty} dt_2 \{g_2(\boldsymbol{t})\} \\ &= e^{-\lambda_1 t_1 - \lambda_2 t_2 - \lambda_{12} \max(t_1, t_2)} - (\lambda_{12}/\lambda_+) e^{-\lambda_+ \max(t_1, t_2)} \qquad \text{for } t_1 > t_2.\end{aligned}$$

This formula, derived for the case $t_1 > t_2$, has been written in a symmetric form which also holds for $t_1 < t_2$. The total probability mass of the continuous component is $\bar{G}^c(\mathbf{0}) = 1 - \lambda_{12}/\lambda_+$, and the singular component is given by

$$\bar{G}^s(\boldsymbol{t}) = \bar{G}(\boldsymbol{t}) - \bar{G}^c(\boldsymbol{t}) = (\lambda_{12}/\lambda_+)e^{-\lambda_+ \max(t_1, t_2)}.$$

$\bar{G}^s(\boldsymbol{t})$ is the probability content of that part of the line $t_1 = t_2$ which lies north east of the point $\boldsymbol{t}$.

In the foregoing, a physical situation was considered and the resulting joint distribution of failure times was derived and investigated. We now show how Theorem 7.3 can be used to reverse the argument, i.e., derive a physical model from the joint distribution. Thus, we start with

$$\bar{G}(\boldsymbol{t}) = \exp\{-\lambda_1 t_1 - \lambda_2 t_2 - \lambda_{12}\max(t_1, t_2)\}.$$

First we verify the fact that the distribution defined by $\bar{G}(\boldsymbol{t})$ allows ties. This can be established as follows: (i) calculate the continuous and singular components of $\bar{G}(\boldsymbol{t})$, as demonstrated above; (ii) note that any singular component must be concentrated on the subspace $\{\boldsymbol{t}: t_1 = t_2\}$ of $(0, \infty)^2$, by inspection of $\bar{G}(\boldsymbol{t})$; (iii) calculate that the total probability mass of the singular component is λ_{12}/λ_+, where $\lambda_+ = \lambda_1 + \lambda_2 + \lambda_{12}$. Thus, $\text{pr}(T_1 = T_2) = \lambda_{12}/\lambda_+$.

Next, we calculate the sub-survivor functions $\bar{F}(c, t)$ for $c = \{1\}, \{2\}$ and $\{1, 2\}$. We have

$$\bar{F}(\{1\}, t) = \text{pr}(t < T_1 < T_2) = \int_{t < t_1 < t_2} g_2(\boldsymbol{t})d\boldsymbol{t},$$

where $g_2(\boldsymbol{t})$ is the density component on the set $\{\boldsymbol{t}: t_1 < t_2\}$ derived explicitly above. Thus,

$$\begin{aligned}\bar{F}(\{1\}, t) &= \int_t^\infty dt_2 \int_t^{t_2} dt_1 \{\lambda_1(\lambda_2 + \lambda_{12})\exp(-\lambda_1 t_1 - \lambda_2 t_2 - \lambda_{12} t_2)\} \\ &= (\lambda_1/\lambda_+)e^{-\lambda_+ t}.\end{aligned}$$

Likewise, $\bar{F}(\{2\}, t) = (\lambda_2/\lambda_+)e^{-\lambda_+ t}$. Lastly,

$$\bar{F}(\{1, 2\}, t) = \bar{G}^s(t\mathbf{1}) = (\lambda_{12}/\lambda_+)e^{-\lambda_+ t},$$

where $\bar{G}^s(\boldsymbol{t})$ is the singular component. For a check on these calculations note that

$$\bar{F}(t) = \sum_c \bar{F}(c, t) = e^{-\lambda_+ t} = \bar{G}(t\mathbf{1}).$$

Finally, note that none of the sub-distributions has a discontinuity so Theorem 7.3 applies in simplified form. The sub-survivor functions can be plugged into the formula there for $\mathrm{pr}(R_C > t)$ to find the probability structure of the $\boldsymbol{R}$-system; the product term in $\mathrm{pr}(R_C > t)$ does not come into play. Thus,

$$\mathrm{pr}(R_{\{1\}} > t) = \exp\left\{-\int_0^t (\lambda_1 e^{-\lambda_+ s}/e^{-\lambda_+ s})ds\right\} = e^{-\lambda_1 t};$$

likewise, $\mathrm{pr}(R_{\{2\}} > t) = e^{-\lambda_2 t}$ and $\mathrm{pr}(R_{\{1,2\}} > t) = e^{-\lambda_{12} t}$. In this way a characterization of the Marshall-Olkin distribution, namely the one based on the three types of fatal shock with exponential waiting times, has been reconstructed as an $\boldsymbol{R}$-system from the form of $\bar{G}(\boldsymbol{t})$ alone.

The sub-hazard function for configuration $c = \{j\}$ is the constant $h(j, t) = \lambda_j$ ($j = 1, 2$). This accords with the way in which the distribution is set up. Note, however, that there is an extra sub-hazard here: $h(\{1, 2\}, t) = \lambda_{12}$ for failure configuration $\{1, 2\}$.

Generalizations of the system, to multivariate and Weibull versions, have been developed by Marshall and Olkin (1967b), Lee and Thompson (1974), David (1974), Moeschberger (1974), and Proschan and Sullo (1974). ■

7.3 Specified marginals

An independent-risks proxy model will reproduce the $\bar{F}(c, t)$ or any given $\bar{G}(\boldsymbol{t})$ but not in general the univariate marginals $\bar{G}_j(t)$, as seen in Example 7.1. Hence, if only we could observe the T_j, separately and unmodified, as well as (C, T), we could eventually identify the $\bar{G}_j(t)$ in addition to the $\bar{F}(c, t)$ and thus detect that the independent-risks proxy model was not the true one. For instance, in controlled experimental situations it might be possible to run different components of a system separately but in joint-operation mode. This is more likely to be relevant to physical, engineering applications than to biological ones.

The marginal survivor functions $\bar{G}_j^*(t)$ of the proxy model may be contrasted with the marginals $\bar{G}_j(t)$ of $\bar{G}(\boldsymbol{t})$. Thus, $\bar{G}_j^*(t)$ and $\bar{G}_j(t)$ both have form $\exp\{-\int_0^t r(s)ds\}$, but with $r(s) = h(j, s)$ for $\bar{G}_j^*(t)$ and $r(s) = h_j(s)$ for $\bar{G}_j(t)$; $h_j(s)$ here is the marginal hazard function of T_j. The coincidence $\bar{G}_j^*(t) = \bar{G}_j(t)$ is thus equivalent to $h(j, t) = h_j(t)$, the Makeham assumption (Theorem 3.3).

Evidently, if the Makeham assumption does not hold there is no independent-risks proxy model that will reproduce both the $\bar{F}(c, t)$ and the $\bar{G}_j(t)$. But is there a dependent-risks proxy model that will do the job? If so, we still have an identifiability problem, though perhaps not one involving independent risks. The following theorem shows that, yes, we still have the problem.

Theorem 7.5 (Crowder, 1991)

Let a set of univariate marginals $\bar{G}_j(t)$ $(j = 1, \ldots, p)$ and a set of sub-survivor functions $\bar{F}(c, t)$ be specified. Suppose that one can find a joint survivor function $\bar{G}(\boldsymbol{t})$ for which (i) the specified functions are matched, and (ii) there exists some set Ω, open in R^p, containing no open subsets of zero probability. Then there exist infinitely many different joint survivor functions satisfying (i) and (ii).

Proof

See the paper cited. ■

The technical proof of this theorem is based on a very simple construction (Lemma 7.6). In Figure 7.1, which illustrates the two-dimensional case, suppose that *A*, *B*, *C*, and *D* are small neighbourhoods within Ω each containing probability mass at least ε. Then one can shift probability mass $\varepsilon/2$ from *A* to *B*, and the same amount from *D* to *C*. The point is that none of the $\bar{F}(j, t)$ or $\bar{G}_j(t)$ $(j = 1, 2)$ is changed. (Refer back to Figure 3.1.) For every such shift we get a different $\bar{G}(\boldsymbol{t})$ with the same sub- and marginal survivor functions.

Condition (ii) of the theorem will be satisfied whenever $\bar{G}(\boldsymbol{t})$ has a positive density over Ω, plus perhaps some atoms of probability or a singular component of positive probability within a subspace. For the result to be non-empty, we have to show that a $\bar{G}(\boldsymbol{t})$ satisfying (i) and (ii) exists. This is done, subject to some coherency conditions on $\bar{G}(\boldsymbol{t})$, in the paper cited: the argument first uses Theorem 7.3, to set up an independent-risks proxy model that matches the sub-survivor functions, and then progressively

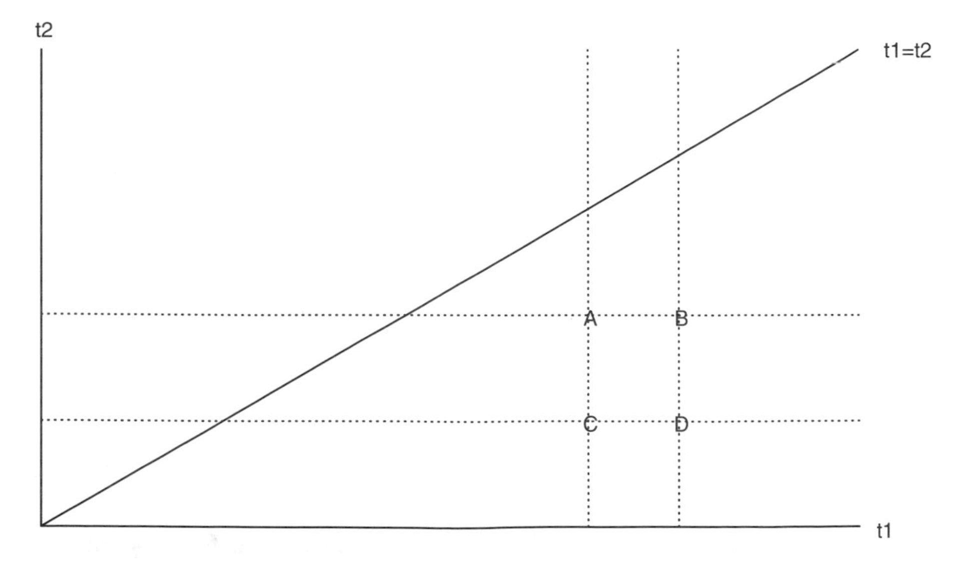

Figure 7.1 Construction for Theorem 7.5.

distorts this joint distribution as described above so that eventually the marginals are also matched.

We can carve up the set of p-dimensional survival distributions into classes whose members have the same sub-survivor functions. One member will be the independent-risks proxy model for the whole class. The class can be further divided into subclasses with the same marginals. Unlimited observation of (C, T) will eventually identify the class, and further unlimited observation of the individual T_j, if possible, will identify the subclass. One will then know whether the data can support independent risks or not.

Suppose now that, failing complete identification of $\bar{G}(\boldsymbol{t})$, we ask about the parametric form of $\bar{G}(\boldsymbol{t})$. In some applications it has been argued that the failure times must have distributions of certain parametric form, in particular, Weibull: e.g., see Pike (1966) and Peto and Lee (1973) in connection with animal survival times under continuous exposure to a carcinogen and Galambos (1978, Section 3.12) in connection with the breakdown of systems and the strength of materials. Unfortunately, the theorem shows that even if the $\bar{F}(c, t)$ are given and parametric forms for the $\bar{G}_j(t)$ are specified, $\bar{G}(\boldsymbol{t})$ is still not identified. The use of a continuous deformation version of the construction in Theorem 7.5 will continuously move the functional form of $\bar{G}(\boldsymbol{t})$ outside any given parametric class without affecting the $\bar{F}(c, t)$ or $\bar{G}_j(t)$.

Lemma 7.6

This is a corrected version of Lemma 1 of Crowder (1991) with a modified proof that allows properly for tied failure times. It applies equally to discrete, continuous and mixed models.

Let a set of univariate marginals $\bar{F}_j(u)$ ($j = 1, \ldots, p$) and a set of sub-survivor functions $\bar{F}(c, u)$ be specified that satisfy

(i) $\sum_{c \ni k} \{\bar{F}(c, u_1) - \bar{F}(c, u_2)\} \le \bar{F}_k(u_1) - \bar{F}_k(u_2)$ for each k and $0 < u_1 < u_2$,

and

(ii) $\sum_c \bar{F}(c, u) \le \bar{F}_k(u)$ for each k and $u > 0$.

Let $\bar{G}(\boldsymbol{t})$ be a joint survivor function with univariate marginals $\bar{G}_j(t_j)$ ($j = 1, \ldots, p$) and with sub-survivor functions equal to the specified $\bar{F}(c, u)$. Let $u_1 < u_2 < u_3$ and suppose that, for some k, $\bar{G}_k(u)$ matches the specified $\bar{F}_k(u)$ at u_1 and u_3. Then additional matching of $\bar{G}_k(u)$ to $\bar{F}_k(u)$ at u_2 can be achieved by a modification of $\bar{G}$ that leaves its other univariate marginals and all of its sub-survivor functions unaltered.

> ***Note.*** Conditions (i) and (ii), expressed in terms of failure time components, are (i) $\text{pr}\{u_1 < T_k \le u_2, T_k = \min_j(T_j)\} \le \text{pr}(u_1 < T_k \le u_2)$, and (ii) $\text{pr}\{\min_l(T_l) > u\} \le \text{pr}(T_k > u)$. These are just requirements for coherency of the joint distribution.

Proof

First, the following sets are defined within R^p: $M_k = \{t\colon t_k = \min_l(t_l)\}$, interpreted as including joint minima where t_k is tied with other components; $A_k = \{t\colon u_1 < t_k \le u_2\}$; $B_k = \{t\colon u_2 < t_k \le u_3\}$; $C_k = \{t\colon u_3 < t_k\}$; $N_k = \{t\colon \min_{l\neq k}(t_l) \le u_2\}$; $\overline{M}_k$ and $\overline{N}_k$ are the respective complements in R^p. (See Figure 7.2.) Next, note that transferring probability mass between points within $\overline{M}_k$ will leave $\bar{G}(c, u)$ unaltered for all $c \ni k$, and transfer along the t_k coordinate direction will leave all $\bar{G}(c, u)$ with $c \ni k$ and all $\bar{G}_l(u)$ for $l \neq k$ unaltered.

Let $\varepsilon = \bar{G}_k(u_2) - \bar{F}_k(u_2)$ be the departure from specification at u_2. Matching of $\bar{G}_k$ to $\bar{F}_k$ at u_2 will be achieved by transferring probability mass ε from the set $B_k \cap \overline{M}_k$ to $A_k \cap \overline{M}_k$; if $\varepsilon < 0$, the transfer is of mass $-\varepsilon$ in the other direction. Provided that the transfer can be made along the t_k-direction, the $\bar{G}$-marginals other than $\bar{G}_k$, and its sub-survivor functions $\bar{G}(c, u)$, will all be undisturbed. It remains to check that there is probability mass $|\varepsilon|$ available under $\bar{G}$ for such a transfer.

If $\varepsilon < 0$, then $A_k \cap \overline{M}_k$ initially needs to contain probability mass $\geq -\varepsilon$ under $\bar{G}$, i.e., it is required that

$$
\begin{aligned}
-\varepsilon \le \bar{G}(A_k \cap \overline{M}_k) &= \bar{G}(A_k) - \bar{G}(A_k \cap M_k) \\
&= \{\bar{G}_k(u_1) - \bar{G}_k(u_2)\} - \sum_{c \ni k} \{\bar{G}(c, u_1) - \bar{G}(c, u_2)\}.
\end{aligned}
$$

Using the definition of ε, together with the matching already present between $\bar{G}$ and $\bar{F}$, this reduces to

$$
0 \le \{\bar{F}_k(u_1) - \bar{F}_k(u_2)\} - \sum_{c \ni k} \{\bar{F}(c, u_1) - \bar{F}(c, u_2)\},
$$

which is condition (i).

The case $\varepsilon > 0$ can be similarly dealt with to show that $\bar{G}(B_k \cap \overline{M}_k) \geq \varepsilon$. However, in this case, the transfer along t_k must be from subset $B_k \cap \overline{M}_k \cap N_k$ of $B_k \cap \overline{M}_k$. This is because $t_k < u_2$ in the destination set $A_k \cap \overline{M}_k$ and the transfer must be performed within $\overline{M}_k$ in which t_k is not the minimum component. If

$$
\eta = \varepsilon - \bar{G}(B_k \cap \overline{M}_k \cap N_k) < 0
$$

to begin with, the problem can be solved by transferring mass η from $B_k \cap \overline{M}_k \cap \overline{N}_k$ to $C_k \cap \overline{M}_k \cap \overline{N}_k$, and then an equal mass from $C_k \cap \overline{M}_k \cap N_k$ to $B_k \cap \overline{M}_k \cap N_k$. This will not disturb any of the specifications, and it remains only to check on the availability of mass η initially in $C_k \cap \overline{M}_k \cap N_k$. Thus, we require

$$
\varepsilon - \bar{G}(B_k \cap \overline{M}_k \cap N_k) \le \bar{G}(C_k \cap \overline{M}_k \cap N_k),
$$

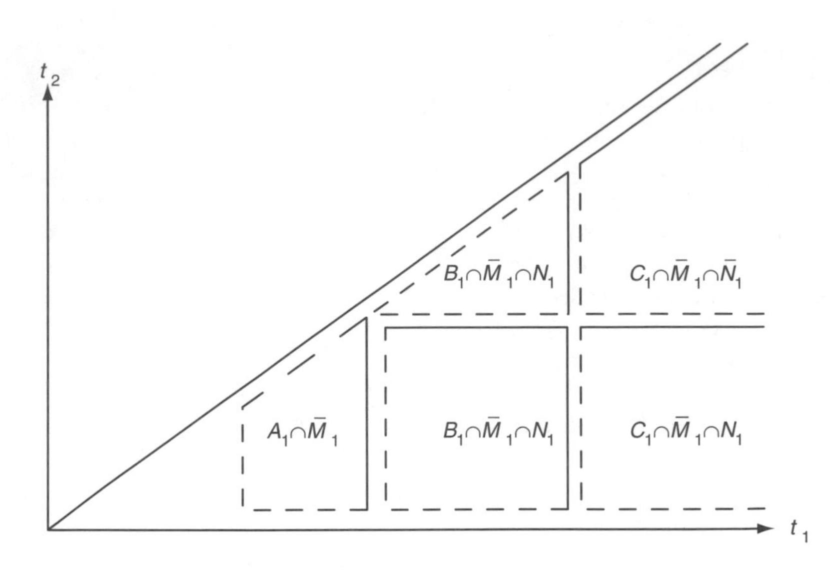

Figure 7.2 Illustration of the construction in Lemma 7.6 for the case $p = 2$, $k = 1$: dashed lines indicate open boundaries of sets, otherwise closed.

i.e.,

$$\begin{aligned}
\varepsilon &\leq \bar{G}\{(B_k \cup C_k) \cap \bar{M}_k \cap N_k\} \\
&= \bar{G}(\{t_k > u_2\} \cap \{t_k \neq \min_l(t_l)\} \cap \{\min_{l \neq k}(t_l) \leq u_2\}) \\
&= \bar{G}(\{t_k > u_2\} \cap \{\min_{l \neq k}(t_l) \leq u_2\}) = \bar{G}(\{t_k > u_2\}) - \bar{G}(\{\min_l(t_l) > u_2\}) \\
&= \bar{G}_k(u_2) - \sum\nolimits_c \bar{G}(c, u_2).
\end{aligned}$$

As before, using the definition of ε and the matching of $\bar{G}$ to $\bar{F}$ already present, this reduces to

$$0 \leq \bar{F}_k(u_2) - \sum\nolimits_c \bar{F}(c, u_2),$$

which is just condition (ii). ■

7.4 Discrete failure times

We show in this section that, in the case of discrete failure times, some assessment of dependence between risks can be made, in contrast to the continuous-time case. The possible failure times are $0 = \tau_0 < \tau_1 < \ldots < \tau_m$ and the quantities $\bar{h}(c, t)$ (Section 5.2) will make a reappearance by popular request. First, a lemma will be found useful.

Lemma 7.7

Let $f^{(1)}(c, t)$ and $f^{(2)}(c, t)$ be the sets of sub-density functions for two discrete-time models that have the same set of configurations. Then the following conditions are equivalent:

(i) $f^{(1)}(c, t) = f^{(2)}(c, t)$ for all (c, t);

(ii) $\bar{h}^{(1)}(c, t) = \bar{h}^{(2)}(c, t)$ for all (c, t);

(iii) $h^{(1)}(c, t) = h^{(2)}(c, t)$ for all (c, t).

Proof

From (i), summing over c, $f^{(1)}(t) = f^{(2)}(t)$. Therefore, $\bar{F}^{(1)}(t) = \bar{F}^{(2)}(t)$ and (ii) follows. Conversely, summing (ii) over c yields

$$f^{(1)}(t)/\bar{F}^{(1)}(t) = f^{(2)}(t)/\bar{F}^{(2)}(t),$$

which, using $f^{(k)}(t) = \bar{F}^{(k)}(t-) - \bar{F}^{(k)}(t)$, is equivalent to

$$\bar{F}^{(1)}(t-)/\bar{F}^{(1)}(t) = \bar{F}^{(2)}(t-)/\bar{F}^{(2)}(t).$$

It follows, by setting $t = \tau_1, \tau_2, \ldots$ in turn, that $\bar{F}^{(1)}(t) = \bar{F}^{(2)}(t)$ for all t. Hence, from (ii), $f^{(1)}(c, t) = f^{(2)}(c, t)$ and so (ii) $\Rightarrow$ (i). That (i) $\Leftrightarrow$ (iii) can be proved similarly. ■

The result now to be given concerns the existence of an independent-risks proxy model of the basic type appearing in Theorem 7.1. That is, there exists $\bar{G}^*(\boldsymbol{t})$, of form $\Pi_{j=1}^{p}\bar{G}_j^*(t_j)$, which reproduces the sub-densities, i.e., such that $g^*(c, t) = f(c, t)$ for all (c, t). The point is that the basic proxy model, which has only p components, reproduces not only the $f(j, t)$ but also all the $f(c, t)$ for non-simple configurations. Not surprisingly, this is only possible when the original system is of fairly special type, actually, one that obeys (iii) of Theorem 5.1.

Theorem 7.8

An independent-risks proxy model of basic type exists for $\bar{G}(\boldsymbol{t})$ if and only if

$$\bar{h}(c, t) = \Pi_{j \in c}\bar{h}(j, t) \qquad \text{for all } (c, t). \tag{7.4.1}$$

In this case, the proxy marginal survivor functions are given uniquely by

$$\bar{G}_j^*(t) = \Pi_{s=1}^{l(t)}\{1 + \bar{h}(j, \tau_s)\}^{-1}, \text{ where } l(t) = \max\{l : \tau_l \le t\}. \tag{7.4.2}$$

Proof

From Lemma 7.3 the proxy assertion, $f(c, t) = g^*(c, t)$ for all (c, t), holds if and only if $\bar{h}(c, t) = \bar{h}^*(c, t)$ for all (c, t). Since the proxy model has independent risks, this in turn implies Equation 7.4.1 by Theorem 5.1 (iii). Conversely, take an independent-risks model with $\bar{h}$-functions $\bar{h}_j^*(t)$ equal to $\bar{h}(j, t)$ for $j = 1, \ldots, p$. Then, Equation 7.4.1 implies that

$$\bar{h}(c, t) = \Pi_{j \in c} \bar{h}_j^*(t),$$

which equals $\bar{h}^*(c, t)$ by Theorem 5.1 (ii), i.e., the proxy assertion holds. Finally, the only possible choice for $\bar{h}_j^*(t)$ has been shown to be $\bar{h}_j(j, t)$. Hence,

$$\bar{h}(j, t) = \bar{h}_j^*(t) = \{\bar{G}_j^*(t-)/\bar{G}_j^*(t)\} - 1,$$

which yields

$$\bar{G}_j^*(t) = \{1 + \bar{h}(j, t)\}^{-1} \bar{G}_j^*(t-),$$

and hence Equation 7.4.2 by iteration. ■

The basic proxy model is constructed by taking its marginal $\bar{h}_j^*(t)$ functions to be the original $\bar{h}(j, t)$. We are whistling the same old Cox-Tsiatis tune (Theorem 7.1), but now more discretely. A couple of examples are given below of dependent-risks models of which one satisfies Equation 7.4.1 and the other does not.

At first sight there seems to be a conflict between the present theorem and the earlier Theorem 7.3. The condition there, that the sub-distributions have no common discontinuities, is here violated. In fact, it would take a rather contrived discrete-time system not to violate the condition. In the discussion after Theorem 7.3 it was shown why the condition was necessary for the situation covered there. A re-examination of that discussion shows why the condition is not necessary for the situation covered here. Basically, we do not have an $R_{\{1,2\}}$ now, only R_js. Thus, there is no question of trying to match $\bar{F}(\{1, 2\}, t)$ by anything but the term arising from chance coincidence of R_1 and R_2. This argument also rules out the possibility of non-basic independent-risks proxy models, i.e., ones with R_c components where c is not a simple index, unless the rather unlikely condition on the discontinuities is met. Thus, for discrete-time systems not subject to the condition, Theorem 7.8 exhausts the possibilities.

We can also generate discrete analogues of the results in Section 7.3. These concern proxy models that match the marginals of $\bar{G}(t)$ as well as its sub-distributions. First, Theorem 7.5 is easily dealt with for the discrete case. Examination of the proof of this theorem, given in Crowder (1991), shows that all we need here is the existence of some k with positive probability at points A, B, C, and D within the set $\{\boldsymbol{t}: t_k < \min_{l \neq k}(t_l)\}$ such that the lines AB and DC are orthogonal to the t_k axis, and AD and BC are parallel to it. This condition will be met, for instance, by any model which puts positive probability on all points in a lattice set $(\tau_l, \tau_{l+3})^p$ for some integer l. Under the condition, and given the existence of a $\bar{G}(\boldsymbol{t})$ that matches the given sub-survivor functions and marginals, the theorem says that there are infinitely many such joint survivor functions.

Example 7.5 Discrete version of Freund (1961)

Referring back to Example 5.3, the vital condition 7.4.1 does hold for this model. Thus, there does exist an independent-risks proxy model, and this is just composed of the two independent sequences of Bernoulli trials used to set up the model. From Equation 7.4.2, the explicit form of the proxy survivor functions is $\bar{G}_j^*(t) = \pi_j^t$. ■

Example 7.6 Discrete version of Marshall-Olkin (1967)

Referring back to Example 5.4, the condition 7.4.1 fails in this case. Thus, there is no independent-risks proxy model with two latent failure times (though there is one with three, as is clear from the way in which the model is set up). ■

7.5 *Regression case*

In view of the results on identifiability of $\bar{G}(\boldsymbol{t})$ from observation of (C, T) it might be thought that, in the continuous-time case, one can never obtain hard information about the distributions of individual component failure times or on the dependence structure between them. However, the foregoing results are restricted to the independent, identically distributed case. It turns out that, when there are explanatory variables in the model, identification is possible within a certain framework.

Theorem 7.9 (Heckman and Honore, 1989)

Suppose that $\bar{G}(\boldsymbol{t};\boldsymbol{x})$ has the form $K(\boldsymbol{v})$, where $\boldsymbol{v} = (v_1, \ldots, v_p)$ with $v_j = \exp\{-\xi_j(\boldsymbol{x})H_j(t_j)\}$, and assume that:

(i) $\partial K/\partial v_j > 0$ and continuous on $\{\boldsymbol{v}: 0 < v_j \leq 1; j = 1, \ldots, p\}$;
(ii) $H_j(t) \to 0$ as $t \to 0$, $H_j(t_0) = 1$ for some t_0, $H_j'(t) = dH_j(t)/dt > 0$ for all t;
(iii) $\xi_j(\boldsymbol{x}_0) = 1$ for some $\boldsymbol{x}_0$, $\{\xi_1(\boldsymbol{x}), \ldots, \xi_p(\boldsymbol{x})\}$ covers the range $(0, \infty)^p$ as $\boldsymbol{x}$ varies.

Then the set of sub-survivor functions $\{\bar{F}(j, t;\boldsymbol{x}): j = 1, \ldots, p\}$ determines the joint survivor function $\bar{G}(\boldsymbol{t};\boldsymbol{x})$.

Proof

From Theorem 3.1,

$$f(j,t;x) = [-\partial K/\partial t_j]_{t\mathbf{1}_p} = \xi_j(x)H_j'(t)\exp\{-\xi_j(x)H_j(t)\}K_j\{v_t(x)\},$$

where $v_t(x)$ has jth component $\exp\{-\xi_j(x)H_j(t)\}$ and $K_j = \partial K/\partial v_j$.

(a)

Consider the ratio

$$\begin{aligned} & f(j,t;x)/f(j,t;x_0) \\ & = \{\xi_j(x)/\xi_j(x_0)\}\exp[-H_j(t)\{\xi_j(x)-\xi_j(x_0)\}][K_j\{v_t(x)\}/K_j\{v_t(x_0)\}]. \end{aligned}$$

Let $t \to 0$. Then $H_j(t) \to 0$, $v_t(x) \to \mathbf{1}_p$ for all x, and the ratio $\to \xi_j(x)/\xi_j(x_0) = \xi_j(x)$. Thus, $\xi_j(x)$ is identified as a function of x from knowledge of $f(j, t;x)$.

(b)

Set $t = t_0$ and let $\{\xi_1(x), \ldots, \xi_p(x)\}$ range over $(0, \infty)^p$. Then K is identified as a function of its arguments from

$$\Sigma\bar{F}(j,t_0;x) = \bar{G}(t_0\mathbf{1}_p;x) = K\{e^{-\xi_1(x)}, \ldots, e^{-\xi_p(x)}\}.$$

(c)

Fix $\xi_k(x)$ and let the other $\xi_j(x)$ all tend to zero. Then $\bar{G}(t\mathbf{1}_p;x) \to K(v_k)$, where v_k has kth element $e^{-\xi_k(x)H_k(t)}$ with the rest all equal to 1. Thus, $\bar{G}(t\mathbf{1}_p;x)$ is a known function of $\xi_k(x)H_k(t)$ alone, known because K has already been identified. Now $H_k(t)$ is identified because $\xi_k(x)$ has already been identified. ■

Heckman and Honore introduced their theorem via the univariate proportional hazards model which has survivor function $\bar{F}(t;x) = e^{-\xi(x)H(t)}$, $H(t)$ being the integrated hazard function; this is where the argument $v_j = e^{-\xi_j(x)H_j(t_j)}$ in K comes from. In the theorem T_j has marginal survivor function $K_j\{e^{-\xi_j(x)H_j(t_j)}\}$, where K_j is the jth marginal of K; this has proportional hazards form if and only if $K(y) = y^\gamma$ for some γ.

The assumptions $\xi_j(x_0) = 1$ and $H_j(t_0) = 1$ are just normalizations. If $\xi_j(x_0)H_j(t_0) = a_j$ in the original K then it can be replaced by K^* defined by $K^*(\ldots v_j \ldots) = K(\ldots v_j^{a_j} \ldots)$. The assumption in (iii), that the $\xi_j(x)$ can vary independently over $(0, \infty)$, is satisfied, for example, by the standard log-linear model $\log\xi_j(x) = x^{\mathrm{T}}\beta_j$ provided that x is unbounded and the vectors β_j can vary independently.

Example 7.1 (continued)

A regression version of Example 7.1 can be obtained by first replacing ν by $\nu\lambda_1\lambda_2$ (with $0 \le \nu \le 1$ now) and then λ_j by $\exp(\alpha_j + x^T\beta_j)$ for $j = 1, 2$. Take $\xi_j(x) = \exp(x^T\beta_j)$, $x_0 = \mathbf{0}$, and $H_j(t) = t$ with $t_0 = 1$. Then $v_j = \exp(e^{-\alpha_j}\lambda_j t_j)$ and

$$\begin{aligned}\bar{G}(t) &= \exp(-\lambda_1 t_1 - \lambda_2 t_2 - \nu\lambda_1 t_1 \lambda_2 t_2) \\ &= v_1^{-a_1} v_2^{-a_2} \exp[-\nu\{\log(v_1^{-a_1})\}\{\log(v_2^{-a_2})\}],\end{aligned}$$

where $a_j = e^{-\alpha_j}$ for $j = 1, 2$. This is the $K(v)$ of the theorem. ■

Heckman and Honore showed that their theorem goes some way towards covering models of the accelerated life type, as well as those of the proportional hazards type at which it is primarily targeted. Such models have univariate survivor functions of form $\bar{F}(t, x) = e^{-H\{t\xi(x)\}}$. This leads to consideration of the multivariate form $\bar{G}(t, x) = K(v)$ with $v_j = \exp[-H_j\{t_j\xi_j(x)\}]$. But v_j can be re-expressed as $\exp[-H_j\{-\log e^{-t_j\xi_j(x)}\}]$, and so $K(v)$ can be re-expressed as $K^*(v^*)$ with $v_j^* = e^{-t_j\xi_j(x)}$, as required for the theorem. Thus K^* and the ξ_j will be identified. Further identification of K and the H_j depends on knowing the marginal distributions of K.

Theorem 7.9 is more deductive than seductive. Although its proof follows the usual logical process, it reveals a possible weakness for practical statistical applications. For instance, in the case of univariate simple random sampling, one can be persuaded that the histogram will begin to give a reasonable facsimile of the underlying density after a moderate number of observations has been recorded. On the other hand, an examination of the proof of Theorem 7.6 suggests that x might have to cover an awful lot of ground before any recognisable picture of the various functional forms emerges.

7.6 *Censoring of survival data*

In standard Survival Analysis, the variate of interest is T_1, the failure time, which is observed unless preceded by censoring, say at time T_2. In general terms this comes naturally under the Competing Risks umbrella with $p = 2$. The classic problem in this context is whether the censoring inhibits inference about the T_1-distribution. It is thus precisely the problem of identifiability dealt with in this chapter.

We have seen that, in the independent, identically distributed case, the marginal distribution of T_1 is not identifiable in the nonparametric sense; nor is that of T_2 but that is of secondary interest. (In the regression case, this total lack of identifiability is somewhat ameliorated, see Section 7.5.) In order to achieve identifiability, some knowledge or assumption, external to the data, has to be provided. A very common case is that of fixed-time censoring, where the T_2 values can be regarded as having been fixed in advance of the observation process. More generally, T_2 might be determined independently of T_1 by some random mechanism, so that "independent-risks" obtains. Then, as shown in Theorem 3.3, the T_1-distribution (and that of T_2) is identified by the observable sub-distributions. In fact, the theorem also shows that identifiability holds under the weaker, Makeham assumption. Thus, it might be the case that T_1 and T_2 are dependent, perhaps through some external factor or conditions affecting both, but that inference is still possible for T_1 without having to consider T_2. In practice, however, postulation of the Makeham condition might be at best optimistic, since its proper justification would require deeper knowledge about aspects of the underlying stochastic mechanisms than is actually available.

Williams and Lagakos (1977) discussed the problem of identifying G_1, the marginal distribution function of T_1, in the presence of censoring. They proposed a condition, which they called "constant sum," under which G_1 is identified by the likelihood function. Kalbfleisch and MacKay (1979) later showed that the condition is equivalent to $h(1, t) = h_1(t)$, i.e., the Makeham assumption for T_1 only. The point is that the data identify the sub-hazard functions, $h(1, t)$ and $h(2, t)$, through the likelihood, and then the constant sum condition carries this identification on to $h_1(t)$, and so to G_1 through $G_1(t) = 1 - \exp\{-\int_0^t h_1(s)ds\}$. Actually, of course, any assumed relation giving $h_1(t)$ in terms of the sub-hazards would do the job just as well. But that's the snag — you've got to conjure up such a relation from somewhere. Williams and Lagakos showed that the constant sum condition is implied by

$$\text{pr}(T_1 \in N_1 \mid T_2 \in N_2, T_1 > t_2) = \text{pr}(T_1 \in N_1 \mid T_1 > t_2)$$

for $t_1 > t_2$, where $N_j = (t_j, t_j + dt_j)$ for $j = 1, 2$. The equality defines survival "independent of the conditions producing censoring." They noted that such a condition is an ingredient of nearly all existing statistical methods for assessing censored survival data. Thus, it seems that the constant sum condition, or partial Makeham assumption in our terms, is a pretty useful theoretical property, but maybe not so easy to nail down in practice.

We can extend the above straightforwardly to Competing Risks proper: in order to identify a subset of marginal distributions, we just need a Makeham-like assumption for these components.

Finally, identification is meant in the non-parametric sense above. If a parametric model is specified for the latent failure times at the outset, none of the above applies! The identifiability problem is now a completely different one, i.e., of parameters within a parametric family. This situation is discussed briefly below, in Section 7.7. With a parametric model, even though one can isolate a factor comprising the contribution of G_1 to the likelihood, the residual factor might contain some of the parameters of G_1 and so "orthogonality" is not achieved (Section 2.1.1). In this case the censoring is said to be **informative**. However, as pointed out by Kalbfleisch and MacKay (1979), the G_1-factor is a partial likelihood in the sense of Cox (1975) and therefore potentially capable of yielding worthwhile inferences.

7.7 Parametric identifiability

Suppose that we have faith in the specified functional form of $\bar{G}(t)$, perhaps through some theoretical argument, or just through a pragmatic belief that the specification will suffice for the current purpose. Then, given incomplete (c, t)-data, the possibilities are (i) complete identifiability of the joint $\bar{G}(t)$ (as in Example 7.1), (ii) partial identifiability of the parameters (as in Example 7.2), and (iii) complete lack of identifiability (as in Example 7.3 with ϕ set to 1, leaving ψ as the only parameter).

Quite a lot of work has appeared on identifiability within parametric families of distributions. Further published work concerns identifiability within parametric families from T alone, from (C, V), where $V = \max(T_1, \ldots, T_p)$, and from V alone. When the observation is (C, T) it is called the **identified minimum**, and T alone is called the **unidentified minimum**, and likewise for (C, V) and V alone. Whereas T is the primary time variate for Competing Risks, V is that for so-called **Complementary Risks** (Basu and Ghosh, 1980). The distinction can be put in terms of **series** and **parallel systems**: in the former the first component failure is fatal to the whole operation which therefore ceases at time T; in the latter, the operation staggers on until the last component dies, which occurs at time V. An example, not specifically concerned with identifiability, is Gross et al's (1971) maximum likelihood estimation of the Freund (1961) bivariate exponential model from observations of V: in a typical application the component lifetimes are those of two kidneys, and symmetry suggests taking $\lambda_1 = \lambda_2$ and $\mu_1 = \mu_2$ (Example 3.2).

Proschan and Sullo (1974) calculated the likelihood functions for the bivariate Marshall-Olkin distribution under various observational settings including that of Competing Risks. Identifiability is established as a by-product. They also considered a model which combines the features of the Marshall-Olkin and Freund systems.

Anderson and Ghurye (1977) investigated identifiability within a family $\mathcal{G}$ of univariate probability densities $g(t)$ which are continuous and positive for t exceeding some value t_0, and such that, for any two members g_1, g_2 of $\mathcal{G}$, $g_1(t)/g_2(t) \to 0$ or ∞ as $t \to \infty$. Their Theorem 2.1 says that the distribution of $V = \max(T_1, \ldots, T_n)$, where the T_j are independent with densities g_j in $\mathcal{G}$, identifies both n and the g_j. The motivation for this arose from a problem in Econometrics. Their proof rests on an examination of the survivor function of V, $\Pi\bar{G}_j(t)$, which involves all the individual survivor functions $\bar{G}_j$ ($j = 1, \ldots, n$), and whose log-derivative can be made to reveal the differing rates of approach to zero of the components. Anderson and Ghurye gave examples of the theorem, including the normal distribution, and other examples where the asymptotic condition fails but the conclusion holds, and yet others where the maximum does not identify the components. Their Theorem 3.1 extends their Theorem 2.1 to a certain class of bivariate normal distributions.

Basu and Ghosh (1978) considered the bivariate normal distribution. They showed in their Theorem 1 that the distribution of the identified minimum (C, T) identifies the parameters μ and Σ; the authors plugged a gap in the proof of this result previously given by Nadas (1971). Their Theorem 2 says that the unidentified minimum T identifies the numerical values of the components of μ and Σ, but not their ordering. Comparing the two theorems, it seems that the additional C-contribution to the (C, T)-distribution tells you which of the two components is more likely to be the minimum and hence differentiates between them. Theorem 3 of the paper gives a similar result to that of Theorem 1 for the trivariate normal under a certain condition on Σ. The authors also mentioned some other distributions where the parameters might or might not be identified by (C, T) or T alone and opined that these other results are much easier to obtain than for the normal on account of the simple explicit forms for the survivor functions.

Basu and Ghosh (1980) gave a complement of Theorem 2.1 of Anderson and Ghurye (1977): they considered the same family $\mathcal{G}$, this time with t tending to the lower end point of the distribution rather than $+\infty$, and with the minimum T instead of the maximum V. In their second theorem they considered two independent observations, one from each of two distinct gamma distributions, and showed that the distribution of the smaller observation, T, identifies both pairs of gamma parameters up to ordering. Their Theorems 3 and 4 do likewise for the Weibull and Gumbel's (1960) second bivariate exponential, and their Theorem 5 gives a similar result for Gumbel's (1960) first distribution but based on the larger of the pair, V. They also gave a result on partial identification of the Marshall-Olkin distribution from T.

Arnold and Brockett (1983) proved identifiability of the joint distribution of latent failure times from (c, t)-data for a certain bivariate Makeham distribution (see Example 4.5). They also considered a mixture model of type given in Equation (3.2.1), but with proportional hazards: $H_j(t) = \psi_j H_0(t)$. They showed that the joint survivor function $\bar{G}(t)$ is identifiable from (c, t)-data if either H_0 or G is known, in the notation of Example 3.3. This partial identifiability is reminiscent of Heckman and Honore's (1989) extension to the accelerated life model (Section 3.4). Further identifiability results were given for multivariate Pareto, Weibull, and Burr families.

Basu and Klein (1982) gave a survey of some results of the present type.

chapter 8

Martingale counting processes in survival data

8.1 Introduction

The methodology was introduced primarily for "event history analysis" in the mid-1970s and has gained great respect, and wide application, for its power in dealing with the theory in a unified way. The progress of an individual is tracked through time and the instantaneous probabilities of various events that can occur are modelled. The general framework comprises a counting process, recording the events, and an intensity process, governing the instantaneous probabilities. Ordinary Survival Analysis is a very simple case, with just one possible event, failure, terminating the process. Competing Risks is a moderate extension, allowing one of p terminating events. In this sense, the general counting process approach may seem to be a bit of an overkill for our context, but it facilitates awkward problems such as the inclusion of time-dependent covariates. Also, the technical apparatus smoothes the application of the usual statistical methodology, constructing likelihoods and deriving properties of statistics.

It has to be said that the approach is rather off-putting for many statisticians. One opens a book on the subject, sees the terms "counting process theory," "continuous-time martingale," "filtration," "predictable process," "compensator," "stochastic integration," "product integral," etc., and feels the need for an aspirin. At this point, another of these terms, "stopping time," may strike a significant chord with the reader. It is true that a certain amount of investment is required to get into the subject, and it will help if one's training has included a course, or three, on Stochastic Processes. That said, however, the application of the theory can be appreciated at arm's length. It is not absolutely necessary to drown in the technicalities to see how the thing works and how to see how the ideas can be applied.

The historical development of the subject has been very fully set out in the introductory part of the book by Andersen et al. (1993). Therefore, in this chapter we shall not attempt to duplicate this, only to remark that the origins go back to Aalen (1975), with a Competing Risks version in Aalen (1976). Also, the pioneering work of Doob (1940 and 1953, Chapter 7) on martingales should not be forgotten. For readable accounts of the subject see the expository papers by Gill (1984), Borgan (1984), and Andersen and Borgan (1985), and, for comprehensive treatments, see Andersen et al. (1993) and Fleming and Harrington (1991). Williams' (1991) book is also highly recommended as a friendly and enthusiastic introduction to probability and discrete-time martingales.

In order to present the basic ideas underlying the approach outlined in this chapter, we will start back at the beginning and develop the machinery in a very informal way, skating over mathematical technicalities. This basis occupies Sections 8.2 to 8.6. Then, in the following sections, we will just cover the first steps, showing how some simple likelihood functions, estimators, and tests can be constructed and evaluated. The huge amount of methodology that follows on from these humble beginnings will be left for the reader to pursue if sufficiently encouraged by the brief outline here; the books by Andersen et al. (1993) and Fleming and Harrington (1991) provide a wealth of material. Our approach, informal and sketchy, might make some purists feel the need of an aspirin, but they are not part of our target audience.

8.2 *Back to basics: probability spaces and conditional expectation*

The **sample space** Ω is the set of all outcomes under consideration, and subsets of Ω are called **events**. A collection of sets that comprises Ω itself, and the primary events of interest together with unions and complements (and, therefore, intersections too, since the intersection of two sets is the complement of the union of their complements), forms a field. A core concept in probability theory is that of a **σ-field** (sigma field): this is a field in which unions are allowed to extend to a countable number of events. We will use an overbar to denote the complement of a set: $\bar{S} = \Omega - S$. Thus, a σ-field always contains Ω and $\varnothing = \bar{\Omega}$ (the empty set), at the very least.

Example 8.1 Die rolling

Suppose that the die is cast (in gaming, not engineering), and that the outcome of interest is the face-value uppermost when the die comes to rest. With 2 rolls of the die, there are 36 possible outcomes of the form (R_1, R_2), R_1 being the face-value on the first roll and R_2 being that on the second; R_1 and R_2 can each take values 1, 2, ..., 6. Consider events of the form $S_r = \{R_1 + R_2 = r\}$. For instance, S_4 contains the 3 outcomes $\{1, 3\}$, $\{2, 2\}$, and $\{3, 1\}$ of Ω, S_{17} is empty, $\bar{S}_r = \{R_1 + R_2 \neq r\}$, and $S_r \cap S_s = \varnothing$ for $r \neq s$. The non-empty S_r are $S_2, S_3, \ldots, S_{12}$, and these 11 form a **partition** of Ω into disjoint sets. As for the number of distinct sets in the σ-field so generated, I make it $\sum_{j=0}^{11} \binom{11}{j} = 2^{11}$. This is a surprisingly huge number, but that's the nature of σ-fields. Its calculation is simpler when one starts with a partition. ■

Suppose that a σ-field $\mathcal{H}$ is defined on Ω. A **probability measure** $P(.)$ assigns a number between 0 and 1 to each set in $\mathcal{H}$ according to the following rules: $P(\Omega) = 1$, $P(\varnothing) = 0$, and $P(\cup_i A_i) = \Sigma_i P(A_i)$ for any countable collection of disjoint sets A_i in $\mathcal{H}$. The triple $(\Omega, \mathcal{H}, P)$ then constitutes a **probability space**.

A real-valued **random variable** X defined on a sample space Ω assigns a real number to each outcome ω in Ω, so $X(\omega) = x$ for some real x. Thus, X imposes a structure on Ω corresponding to the different values that it takes. If X is discrete, these sets can be taken to be of the form $\{\omega \in \Omega{:}X(\omega) = x\}$, or just $\{X = x\}$, meaning the set of outcomes ω in Ω at which X takes the real value x. Then, X defines a partition of Ω and when we extend this by bringing in all unions and complements, we obtain $\sigma(X)$, the **σ-field generated by X**. If X is continuous, the primary sets are usually taken to be of the form $\{X \le x\}$. The σ-field extension here yields all events of the form $\{X \in A\}$, where A is an interval or union of intervals on the real line.

Suppose that Y is a random variable defined on Ω whose value is determined by that of X, i.e., Y is a function of X. Then, for each y, the subset $\{Y \le y\}$ of Ω belongs to $\sigma(X)$; this is because $\{Y \le y\}$ can be expressed as $\{\omega{:}X(\omega) \in A_y\}$, where A_y is just the set of ω-values in Ω yielding $Y \le y$, and $A_y \in \sigma(X)$. In this case, we say that Y is measurable with respect to $\sigma(X)$. Equivalently, $\sigma(Y) \subset \sigma(X)$, meaning that $\sigma(Y)$ is contained within $\sigma(X)$, i.e., $\sigma(X)$ defines a finer partition of Ω than $\sigma(Y)$, each set of $\sigma(Y)$ being a union of sets of $\sigma(X)$. Conversely, if Y is measurable with respect to $\sigma(X)$, Y is constant on ω-sets on which X is constant and so is a function of X.

Example 8.1 (continued)

Consider the discrete random variable $X = R_1 + R_2$, the sum of the face values after rolling the die twice. The sets in the X-partition of Ω are just the S_r: X takes value r uniquely on S_r, i.e., $S_r = \{\omega \in \Omega{:}X(\omega) = r\}$. The σ-field $\sigma(X)$ can now be generated by repeated application of the set operations, union and complementation, to the S_r and their progeny, as illustrated previously. Take $Y = |X - 3|$, so Y is a function of X that takes values 0, 1, 2, and 3; $\{Y \le 0\} = S_3$, $\{Y \le 1\} = S_3 \cup S_2 \cup S_4$, $\{Y \le 2\} = S_3 \cup S_2 \cup S_4 \cup S_1 \cup S_5$, and $\{Y \le 3\} = \Omega$. So, $\{Y \le y\} \in \sigma(X)$ for each y, and therefore Y is $\sigma(X)$-measurable. ■

More generally, if, for each y, $\{Y \le y\} \in \mathcal{H}$, where $\mathcal{H}$ is a σ-field defined on Ω, we say that Y is measurable with respect to $\mathcal{H}$ or just **$\mathcal{H}$-measurable**; equivalently, $\sigma(Y) \subset \mathcal{H}$. If a probability measure $P(.)$ is defined on $\mathcal{H}$ then, provided that Y is $\mathcal{H}$-measurable, $P(\{Y \le y\})$ is defined because $\{Y \le y\} \in \mathcal{H}$. In this way, the probability of any Y-event expressible in terms of basic $\{Y \le y\}$-sets (by countable unions and complements) is computable.

A central concept in the subsequent development is that of **conditional expectation**. Suppose that X and Y are two random variables defined on the same sample space Ω. We assume that a probability measure $P(.)$ is defined on a σ-field of events that includes the one generated by the sets $\{X \le x, Y \le y\}$. For $A \in \sigma(X)$ let

$$F(y \mid A) = \text{pr}(Y \le y \mid X \in A) = P(Y \le y \cap X \in A) \div P(X \in A).$$

Then the conditional expectation of Y, given $X \in A$, is defined as

$$\text{E}(Y \mid X \in A) = \int y dF(y|A). \tag{8.2.1}$$

If Y is a continuous random variable, $dF(y \mid A)/dy$ is its conditional density function, say $f(y \mid A)$, and then Equation 8.2.1 can be written in the more familiar form $\int yf(y|A)dy$. If Y is discrete, $dF(y \mid A)$ is given by differencing as

$$\text{pr}(Y \le y \mid X \in A) - \text{pr}(Y \le y- \mid X \in A),$$

and then Equation 8.2.1 is evaluated as a sum over the y-values for which $dF(y \mid A) > 0$.

If $A = \{X = x\}$, the conditional expectation Equation 8.2.1 is $\text{E}(Y \mid X = x)$. We can write $\text{E}(Y \mid X)$ to denote the function of X that takes the value $\text{E}(Y \mid X = x)$ at $X = x$. Being a function of X, $\text{E}(Y \mid X)$ is $\sigma(X)$-measurable. Regarded as a function of ω, where $\omega \in \Omega$, $\text{E}(Y \mid X = x)$ is constant over the set $\{\omega \in \Omega: X(\omega) = x\}$. Likewise, $\text{E}(Y \mid X \in A)$ is constant over the set $A \in \sigma(X)$, which is natural because Y has been averaged over A. Again, we can write $\text{E}\{Y \mid \sigma(X)\}$ to mean a function that takes value $\text{E}(Y \mid X \in A)$ when the particular set A in $\sigma(X)$ is specified. Since $\text{E}\{Y \mid \sigma(X)\}$ is constant over the specified set in $\sigma(X)$, it is $\sigma(X)$-measurable. More generally, we can write $\text{E}(Y \mid \mathcal{H})$ for conditioning on an arbitrary σ-field $\mathcal{H}$, and then $\text{E}(Y \mid \mathcal{H})$ is $\mathcal{H}$-measurable. When $\mathcal{H}$ is the minimal σ-field, comprising just Ω and $\varnothing$, $\text{E}(Y \mid \mathcal{H}) = \text{E}(Y)$.

If Y is $\sigma(X)$-measurable, it is a function of X and then its conditional expectation, given $X = x$, is just $Y(x)$, i.e., $\text{E}(Y \mid X = x) = Y(x)$ or just $\text{E}(Y \mid X) = Y$. Suppose now that Z is a third random variable on Ω, not necessarily $\sigma(X)$-measurable. Then $\text{E}(YZ \mid X = x)$ is calculated by averaging YZ over the set $\{X = x\}$. However, since Y is constant over this set, the averaging is effectively only performed over the Z-values, i.e., $\text{E}(YZ \mid X = x) = Y(x)\ \text{E}(Z \mid X = x)$, or just $\text{E}(YZ \mid X) = Y\ \text{E}(Z \mid X)$. More generally, if Y is **$\mathcal{H}$-measurable**, then $\text{E}(Y \mid \mathcal{H}) = Y$ and

$$\text{E}(YZ \mid \mathcal{H}) = Y\, \text{E}(Z \mid \mathcal{H}). \tag{8.2.2}$$

Suppose that $\mathcal{H}_1$ and $\mathcal{H}_2$ are two σ-fields defined on Ω such that $\mathcal{H}_1 \subset \mathcal{H}_2$, i.e., $\mathcal{H}_1$ is a sub-σ-field of $\mathcal{H}_2$. By this is meant that the sets of $\mathcal{H}_1$ are all unions of sets of $\mathcal{H}_2$, i.e., $\mathcal{H}_2$ is a refinement of $\mathcal{H}_1$ defining a finer partition of Ω. If Y is $\mathcal{H}_1$-measurable, then it is also $\mathcal{H}_2$-measurable, because $\{Y \le y\}$'s belonging to $\mathcal{H}_1$ implies its belonging also to $\mathcal{H}_2$. Again, for any Z defined on Ω, $\text{E}(Z \mid \mathcal{H}_1)$ is $\mathcal{H}_1$-measurable, as noted above, and therefore also $\mathcal{H}_2$-measurable by the preceding argument for Y. This fact can be symbolised as

$$\text{E}\{\text{E}(Z \mid \mathcal{H}_1) \mid \mathcal{H}_2\} = \text{E}(Z \mid \mathcal{H}_1). \tag{8.2.3}$$

Conversely, if we average first over $\mathcal{H}_2$, to obtain $E(Z \mid \mathcal{H}_2)$, and then average this over $\mathcal{H}_1$, we obtain the same result as we would have had we performed the coarser averaging over $\mathcal{H}_1$ in the first place:

$$E\{E(Z \mid \mathcal{H}_2) \mid \mathcal{H}_1\} = E(Z \mid \mathcal{H}_1). \tag{8.2.4}$$

To see how this works, informally, take Z discrete and let $B \in \mathcal{H}_1$ with $B = \cup A_i$ for disjoint $A_i \in \mathcal{H}_2$. Then,

$$\begin{aligned} E\{E(Z|\mathcal{H}_2)|B\} &= \sum_i E(Z|A_i)\text{pr}(A_i|B) = \sum_i \{\sum_z z dF(z|A_i)\}\text{pr}(A_i|B) \\ &= \sum_z z\{\sum_i dF(z|A_i)\text{pr}(A_i|B)\} = \sum_z z\{\sum_i dF(z|A_i \cap B)\text{pr}(A_i|B)\} \\ &= \sum_z z\{\sum_i dF(z \cap A_i|B)\} = \sum_z z dF(z|B) = E(Z|B). \end{aligned}$$

In both Equation 8.2.3 and Equation 8.2.4, we end up with the coarser averaging over $\mathcal{H}_1$. If $\mathcal{H}_1 = \{\Omega, \varnothing\}$, the minimal σ-field, then $\mathcal{H}_1$ is contained in any $\mathcal{H}_2$, and then taking $\mathcal{H}_2 = \mathcal{H}$ in Equation 8.2.4 yields

$$E\{E(Z \mid \mathcal{H})\} = E(Z). \tag{8.2.5}$$

Finally, we need one more definition: if $\mathcal{H}_1$ and $\mathcal{H}_2$ are two σ-fields defined on Ω, we will use $\sigma\{\mathcal{H}_1 \cup \mathcal{H}_2\}$ to mean the smallest σ-field containing them both, i.e., the σ-field generated by the sets of $\mathcal{H}_1$ and $\mathcal{H}_2$ together.

8.3 Filtrations

Suppose that we are observing some stochastic process, $\{X(s): s \geq 0\}$, and suppose that $\mathcal{H}_t$ is a partition of the sample space into sets each representing a possible **sample path** of X up to time t. Thus, given a set in $\mathcal{H}_t$, we know the values, i.e., the sample path, taken by $X(s)$ over the time period $[0, t]$. Assume that, as time goes on, no information is lost, so all that is present in $\mathcal{H}_s$ is included in $\mathcal{H}_t$ for $s \leq t$. In other words, as t increases, the partition becomes finer and finer to accommodate the information accruing on the X-values over the extending time period. Now redefine $\mathcal{H}_t$ as the σ-field generated by such a partition, i.e., $\mathcal{H}_t = \sigma\{X(s): 0 \leq s \leq t\}$. Then, for $s < t$, $\mathcal{H}_s \subset \mathcal{H}_t$ and we say that $\{\mathcal{H}_t: t \geq 0\}$, or just $\{\mathcal{H}_t\}$, is an increasing sequence of σ-fields, or **filtration**.

It is possible that $\mathcal{H}_t$ contains more information than the bare histories of the X-process up to time t — it might also record other events along the way, so that $\mathcal{H}_t$ properly contains $\sigma\{X(s): 0 \leq s \leq t\}$ for each t. Nevertheless, so long as $\mathcal{H}_s \subset \mathcal{H}_t$ for $s < t$, $\{\mathcal{H}_t\}$ is a filtration. Provided that $\mathcal{H}_t$ does contain $\sigma\{X(s): 0 \leq s \leq t\}$, the process X is said to be **adapted** to $\{\mathcal{H}_t\}$. Conditional on $\mathcal{H}_t$, i.e., given a particular set of $\mathcal{H}_t$, there is no randomness left in $\{X(s): 0 \leq s \leq t\}$ because the sample path up to time t becomes known and therefore fixed. Conditional on $\mathcal{H}_s$, with $s < t$, there is some randomness left in $\{X(s): 0 \leq s \leq t\}$, namely, the sample path over the additional period $(s, t]$.

Assuming that X is adapted to $\{\mathcal{H}_t\}$, the conditional expectation $E\{X(t) \mid \mathcal{H}_s\}$ is defined for $s < t$ by averaging over the part of the sample path still "at random," i.e., the part over the period $(s, t]$. The given set of $\mathcal{H}_s$, on which we are conditioning, is further partitioned by $\mathcal{H}_t$ into subsets that are assigned probabilities by the process specification, and the averaging is performed with respect to these probabilities. Conversely, for $s \geq t$, $E\{X(t) \mid \mathcal{H}_s\} = X(t)$.

To take an example in discrete time, suppose that X is formed as a cumulative sum, $X(t) = \sum_{i=0}^{t} U_i$ for $t = 0, 1, 2, \ldots$, where the U_i are not necessarily independent or identically distributed. Then, $\mathcal{H}_s = \sigma\{X(r): 0 \leq r \leq s\}$ fixes $U_0, \ldots, U_s$ and leaves $U_{s+1}, \ldots, U_t$ to be generated by the underlying probability law. In consequence,

$$E\{X(t) \mid \mathcal{H}_s\} = \sum_{i=0}^{s} U_i + \sum_{i=s+1}^{t} E\{U_i \mid \mathcal{H}_s\}.$$

8.4 Martingales

8.4.1 Discrete time

Consider a discrete-time stochastic process $\{X(t): t = 0, 1, 2, \ldots\}$ and a filtration $\{\mathcal{H}_t: t = 0, 1, 2, \ldots\}$ that together satisfy the following conditions:

(i) X is adapted to $\{\mathcal{H}_t\}$;
(ii) $E\,|X(t)| < \infty$;
(iii) $E\{X(t) \mid \mathcal{H}_{t-1}\} = X(t-1)$. (8.4.1)

Then X is said to be a **martingale** with respect to $\{\mathcal{H}_t\}$. Since $X(0)$ is $\mathcal{H}_{t-1}$-measurable for $t > 0$, (i), (ii), and (iii) hold equivalently for $X(t) - X(0)$. Therefore, we can take $X(0) = 0$ without loss of generality in describing properties of martingales. If X is a martingale, the **martingale differences** $dX(t) = X(t) - X(t-1)$ satisfy

$$\begin{aligned} E\{dX(t) \mid \mathcal{H}_{t-1}\} &= E\{X(t) \mid \mathcal{H}_{t-1}\} - E\{X(t-1) \mid \mathcal{H}_{t-1}\} \\ &= X(t-1) - X(t-1) = 0. \end{aligned} \tag{8.4.2}$$

Thus, martingales are zero-drift processes adapted to some filtration. Often, this filtration is just $\sigma\{X(0), \ldots, X(t)\}$ itself. For $s < t$, $\mathcal{H}_s \subseteq \mathcal{H}_{t-1}$ so, using Equation 8.2.4,

$$E\{dX(t) \mid \mathcal{H}_s\} = E[E\{dX(t) \mid \mathcal{H}_{t-1}\} \mid \mathcal{H}_s] = E\{0 \mid \mathcal{H}_s\} = 0.$$

It follows that

$$\begin{aligned} E\{X(t) \mid \mathcal{H}_s\} &= E\{X(s) + dX(s+1) + \ldots + dX(t) \mid \mathcal{H}_s\} \\ &= E\{X(s) \mid \mathcal{H}_s\} = X(s). \end{aligned} \tag{8.4.3}$$

In fact, martingales are often more naturally defined as sums, $X(t) = \sum_{s=0}^{t} dX(s)$, in which $E\{dX(s) \mid \mathcal{H}_{s-1}\} = 0$. The $dX(s)$ are zero-mean, uncorrelated increments because, by Equation 8.2.5,

$$E\{dX(s)\} = E[E\{dX(s) \mid \mathcal{H}_{s-1}\}] = E(0) = 0,$$

and, for $s < t$,

$$E\{dX(s)dX(t)\} = E[E\{dX(s)dX(t) \mid \mathcal{H}_{t-1}\}] = E[dX(s)E\{dX(t) \mid \mathcal{H}_{t-1}\}] = E(0) = 0.$$

Example 8.2 Likelihood ratios

Let $U_1, U_2, \ldots,$ be a sequence of independent, identically distributed, continuous variates with common density function $f(u)$. Consider the hypotheses $H_0: f = f_0$ (null) and $H_1: f = f_1$ (alternative). Then the likelihood ratio for testing H_1 vs. H_0, based on observations $(U_1, \ldots, U_n)$, is

$$X(n) = \mathrm{lr}(U_1, \ldots, U_n) = \Pi_{i=1}^{n} \{f_1(U_i)/f_0(U_i)\}.$$

Taking $\mathcal{H}_n = \sigma(U_1, \ldots, U_n)$, X is adapted to $\{\mathcal{H}_n\}$ and, assuming that $E_0 |f_1(U)/f_0(U)| < \infty$ (where E_0 denotes expectation under H_0), X is a martingale with respect to $\{\mathcal{H}_n\}$ under H_0 because

$$\begin{aligned} E_0\{X(n)|\mathcal{H}_{n-1}\} &= X(n-1)E_0\{f_1(U_n)/f_0(U_n)|\mathcal{H}_{n-1}\} \\ &= X(n-1)\int\{f_1(u)/f_0(u)\}f_0(u)du \\ &= X(n-1)\int f_1(u)du = X(n-1). \end{aligned}$$

The martingale differences here are

$$dX(n) = X(n) - X(n-1) = X(n-1)[\{f_1(U_n)/f_0(U_n)\} - 1\}. \blacksquare$$

A process $\{Y(t): t = 0, 1, 2, \ldots\}$ is said to be **predictable** with respect to $\{\mathcal{H}_t: t = 0, 1, 2, \ldots\}$ if $Y(t)$ is $\mathcal{H}_{t-1}$-measurable for each t, i.e., the value of $Y(t)$ is determined by events up to time $t-1$. So, $Y(t)$ can serve as a one-step-ahead prediction of some random variable. For example, if $\mathcal{H}_t$ is $\sigma\{X(0), \ldots, X(t)\}$, $Y(t)$ is just a function of $X(0), \ldots, X(t-1)$. If Z is a stochastic process adapted to $\{\mathcal{H}_t\}$, then we can write

$$dZ(t) = Z(t) - Z(t-1) = dY(t) + dX(t),$$

where

$$dY(t) = E\{dZ(t) \mid \mathcal{H}_{t-1}\},\ dX(t) = dZ(t) - E\{dZ(t) \mid \mathcal{H}_{t-1}\}.$$

Here,

$$Y(t) = Y(0) + \sum_{s=1}^{t} dY(s)$$

is predictable, because $dY(s)$ is $\mathcal{H}_{s-1}$-measurable, and therefore $\mathcal{H}_{t-1}$-measurable, for $s \le t$, and the $dX(t)$ are martingale differences with respect to $\{\mathcal{H}_t\}$, because $E\{dX(t) \mid \mathcal{H}_{t-1}\} = 0$. Thus follows the famous **Doob decomposition**,

$$Z(t) = Y(t) + X(t), \tag{8.4.4}$$

expressing Z as the sum of a predictable process, known as its **compensator**, and a martingale.

8.4.2 Continuous time

A continuous-time stochastic process $\{X(t): t \ge 0\}$ is a martingale with respect to the filtration $\{\mathcal{H}_t: t \ge 0\}$ if, for all $t \ge 0$,

(i) X is adapted to $\{\mathcal{H}_t\}$;
(ii) $E|X(t)| < \infty$;
(iii) $E\{X(t) \mid \mathcal{H}_s\} = X(s)$ for $s \le t$. (8.4.5)

The theory for martingales in continuous time is much more technically sophisticated than that for discrete time because of the possibility of sample path pathology, e.g., discontinuity, and the fact that the time points are uncountable, whereas σ-fields, and the probabilities defined on them, deal only with countable collections of sets. One of the main results is the **Doob-Meyer decomposition**. This states that a continuous-time stochastic process Z adapted to a filtration $\{\mathcal{H}_t\}$ can be expressed in the form

$$Z(t) = Y(t) + X(t), \tag{8.4.6}$$

where $Y(t)$ is predictable, i.e., measurable with respect to $\mathcal{H}_{t-}$, and X is a martingale with respect to $\{\mathcal{H}_t\}$.

The **predictable variation** $\langle X \rangle$ of a martingale X is the compensator of the process $X(t)^2$. The latter has increments

$$\begin{aligned} d\{X(t)^2\} &= X\{(t+dt)-\}^2 - X(t-)^2 = \{X(t-) + dX(t)\}^2 - X(t-)^2 \\ &= dX(t)^2 + 2X(t-)dX(t). \end{aligned}$$

Now,

$$E\{2X(t-)dX(t) \mid \mathcal{H}_{t-}\} = 2X(t-)E\{dX(t) \mid \mathcal{H}_{t-}\} = 0,$$

so, by the definition of $\langle X \rangle$ as the compensator of $X(t)^2$,

$$d <X(t)> = \mathrm{E}[d\{X(t)^2\} \mid \mathcal{H}_{t-}] = \mathrm{E}\{dX(t)^2 \mid \mathcal{H}_{t-}\} = \mathrm{var}\{dX(t) \mid \mathcal{H}_{t-}\}. \quad (8.4.7)$$

In similar vein, the **predictable covariation** $<X_1, X_2>$, where X_1 and X_2 are both martingales with respect to $\mathcal{H}_t$, is the compensator of the process $X_1(t)X_2(t)$. This is defined by the increment $\mathrm{E}\{d(X_1X_2)(t) \mid \mathcal{H}_{t-}\}$. But,

$$\begin{aligned} d(X_1X_2)(t) &= X_1\{(t+dt)-\}X_2\{(t+dt)-\} - X_1(t-)X_2(t-) \\ &= \{X_1(t-)+dX_1(t)\}\{X_2(t-)+dX_2(t)\} - X_1(t-)X_2(t-) \\ &= X_1(t-)dX_2(t) + X_2(t-)dX_1(t) + dX_1(t)dX_2(t). \end{aligned}$$

Hence,

$$d <X_1, X_2> (t) = \mathrm{E}\{d(X_1X_2)(t) \mid \mathcal{H}_{t-}\} = \mathrm{E}\{dX_1(t)dX_2(t) \mid \mathcal{H}_{t-}\} = \mathrm{cov}\{dX_1(t), dX_2(t) \mid \mathcal{H}_{t-}\};$$

X_1 and X_2 are **orthogonal** if $<X_1, X_2> (t) = 0$ for all t. Note that $<X, X> = <X>$ in the notation here.

Let A be predictable and let X be a martingale, both with respect to $\{\mathcal{H}_t\}$. Define the **stochastic integral**

$$Y(t) = \int_0^t A(s)dX(s).$$

Then Y is also a martingale with respect to $\{\mathcal{H}_t\}$ because

$$\mathrm{E}\{dY(t) \mid \mathcal{H}_{t-}\} = \mathrm{E}\{A(t)dX(t) \mid \mathcal{H}_{t-}\} = A(t)\,\mathrm{E}\{dX(t) \mid \mathcal{H}_{t-}\} = 0.$$

Further, the predictable variation increments of Y are

$$d <Y(t)> = \mathrm{var}\{A(t)dX(t) \mid \mathcal{H}_{t-}\} = A(t)^2\, d <X(t)>.$$

Hence, the predictable variation process of Y is given as

$$<Y(t)> = \int_0^t A(s)^2 d<X(s)>. \quad (8.4.8)$$

Likewise, the predictable covariation process of the stochastic integrals $\int_0^t A_j(s)dX_j(s)$ $(j = 1, 2)$ is

$$\int_0^t A_1(s)A_2(s)d<X_1, X_2>(s).$$

8.5 Counting processes

Suppose that $N(t)$ records the number of events of some point process on the time interval $[0, t]$: thus, the increment $dN(t) = N\{(t + dt)-\} - N(t-)$ takes the value 1 if a point event occurs at time t, and 0 if not. Then, N is referred to as a **counting process**. Assume that N is adapted to $\{\mathcal{H}_t\}$. The **intensity function** $\lambda(.)$ of N is essentially a hazard function defined by

$$\text{pr}\{dN(t) = 1 \mid \mathcal{H}_{t-}\} = \lambda(t)dt. \tag{8.5.1}$$

Denote by $\Lambda(t)$ the integrated intensity function,

$$\Lambda(t) = \int_0^t \lambda(s)ds.$$

Then, $d\Lambda(t) = \lambda(t)dt$ and, since $dN(t)$ is just a binary variable, Equation 8.5.1 yields

$$\text{E}\{dN(t) \mid \mathcal{H}_{t-}\} = d\Lambda(t). \tag{8.5.2}$$

From Equation 8.5.2 it can be seen that $\lambda(t)$ and $\Lambda(t)$ are predictable processes, being measurable with respect to $\mathcal{H}_{t-}$. Let $M(t) = N(t) - \Lambda(t)$. Then M is a martingale with respect to $\{\mathcal{H}_t\}$ because

$$\text{E}\{dM(t) \mid \mathcal{H}_{t-}\} = \text{E}\{dN(t) - d\Lambda(t) \mid \mathcal{H}_{t-}\} = d\Lambda(t) - d\Lambda(t) = 0;$$

M is then a **martingale counting process**, and Λ is the compensator of the counting process N.

From Equation 8.4.7, the predictable variation process $\langle M \rangle$ has increments

$$d \langle M(t) \rangle = \text{var}\{dM(t) \mid \mathcal{H}_{t-}\} = \text{var}\{dN(t) \mid \mathcal{H}_{t-}\} = d\Lambda(t)\{1 - d\Lambda(t)\} \approx d\Lambda(t).$$

Thus, $\langle M(t) \rangle = \Lambda(t)$. Further, if A is predictable with respect to $\{\mathcal{H}_t\}$, the stochastic integral $Y(t) = \int_0^t A(s)dM(s)$ is a martingale, and its predictable variation process is, by Equation 8.4.8,

$$\langle Y(t) \rangle = \int_0^t A(s)^2 d\langle M(s) \rangle = \int_0^t A(s)^2 d\Lambda(s). \tag{8.5.3}$$

In a multivariate counting process, $N(t) = \{N_1(t), \ldots, N_p(t)\}$, the components $N_j(t)$ are all counting processes defined on the same sample space. Assume that a filtration $\{\mathcal{H}_t\}$ is defined with respect to which N is adapted and that the intensity process for N_j is Λ_j with respect to $\{\mathcal{H}_t\}$. Let $M_j(t) = N_j(t) - \Lambda_j(t)$ be the jth coordinate martingale counting process. Then, the predictable covariations are given by the increments

$$\begin{aligned} d\langle M_j, M_k \rangle(t) &= \text{cov}\{dM_j(t), dM_k(t) \mid \mathcal{H}_{t-}\} = \text{cov}\{dN_j(t), dN_k(t) \mid \mathcal{H}_{t-}\} \\ &= \text{E}\{dN_j(t)dN_k(t) \mid \mathcal{H}_{t-}\} = \text{pr}\{dN_j(t) = 1 \cap dN_k(t) = 1 \mid \mathcal{H}_{t-}\}, \end{aligned}$$

since $dN_j(t)$ and $dN_k(t)$ are binary variates. We will assume throughout that there cannot be more than one jump at a time, so $dN_j(t)$ and $dN_k(t)$ cannot both be non-zero simultaneously for $j \neq k$. Then, the last expression here is 0, and so M_j and M_k are orthogonal for $j \neq k$.

8.6 Product integrals

The last element in our exciting build-up to the counting process approach in Survival Analysis is the concept of the product integral. This has much theory and many ramifications attached to it — see the seminal paper by Gill and Johansen (1990). We will view it mainly as a notation that facilitates the construction of likelihood functions and the derivation of properties of estimators and tests.

Consider a failure time variate T with survivor function $\bar{F}(t) = \text{pr}(T > t)$ and hazard function $h(t)$. In the discrete-time case, the following expression, obtained by multiplying the successive conditional probabilities together, was given in Section 4.1:

$$\bar{F}(t) = \Pi\{1 - h(\tau_s)\}, \tag{8.6.1}$$

where the product is over $\{s: \tau_s \leq t\}$, the τ_s being the discrete time points. Suppose now that the τ_s define an increasingly fine grid on the time axis and that we can take a limit as their separations tend to zero (Section 4.1.2). Then time becomes continuous and we replace the discrete hazard $h(\tau_s)$, which is a conditional probability, by $h(t)dt$, $h(t)$ being the corresponding continuous-time hazard function, which is a conditional probability density. The right hand side of Equation 8.6.1 becomes $\Pi\{1 - h(s)ds\}$, where the product is now over $\{s: s \leq t\}$, however, it must be said that such a "product," over a continuum of s-values, is only very informally defined at this stage.

We now reframe this by bringing in N, the **counting process associated with T**, which jumps from 0 to 1 at time T. Thus, $N(t) = I\{T \leq t\}$, the indicator function taking value 0 if $t < T$ and 1 if $t \geq T$. If $\mathcal{H}_t$ represents $\sigma\{N(s): 0 \leq s \leq t\}$, the history generated by the N-process, reference to Equation 8.5.1 suggests that we can interpret $h(s)ds$ as $d\Lambda(s)$, by matching λ to h and $\text{pr}\{dN(t) = 1 \mid \mathcal{H}_{t-}\}$ to $\text{pr}(T = t \mid \mathcal{H}_{t-})$. Thus, we can write

$$\bar{F}(t) = \mathcal{P}_{s \leq t}\{1 - d\Lambda(s)\}, \tag{8.6.2}$$

where the **product integral** $\mathcal{P}_{s \leq t}$ is very carefully defined analytically to have sensible mathematical properties: see Gill and Johansen (1990). (A script uppercase pi, when such a font is available, is preferable as a special notation for this operator, ibid.) In the purely continuous case, for instance, Equation 8.6.2 reduces to $\bar{F}(t) = \exp\{-\Lambda(t)\}$ (Section 4.1.2); in the purely discrete case it resumes the form Equation 8.6.1.

8.7 Survival data

The material from here on is a small selection of that covered, in far greater breadth and depth, in Fleming and Harrington (1991) and Andersen et al. (1992). Here, just enough of the basic ideas are set out to serve our purpose.

8.7.1 A single lifetime

Consider first a single failure time T having a continuous distribution on $(0, \infty)$. Let N be the counting process associated with T, $N(t) = \mathrm{I}\{T \leq t\}$. $N(t)$ counts the number of failures up to time t, in this case only 0 or 1 since we are dealing with a single unit. Now let $dN(t)$ be the increment as defined in Section 8.5. Then $dN(t) = 1$ if N jumps at time t, i.e., if $T = t$; otherwise, $dN(t) = 0$. In consequence, taking $\mathcal{H}_t = \sigma\{N(s): 0 \leq s \leq t\}$,

$$\mathrm{E}\{dN(t) \mid \mathcal{H}_{t-}\} = \mathrm{pr}\{T \in [t, t + dt) \mid \mathcal{H}_{t-}\} = Y(t)h(t)dt, \tag{8.7.1}$$

where $h(t)$ is the hazard function of T and

$$Y(t) = \mathrm{I}\{T \geq t\} = 1 - \mathrm{I}\{T < t\} = 1 - N(t-). \tag{8.7.2}$$

If $t > T$, the right hand side of Equation 8.7.1 is zero because then $Y(t) = 0$. In this case, the probability on the left hand side is also zero because the occurrence of the event $\{T < t\}$ will be part of the information carried by $\mathcal{H}_{t-}$, thus ruling out the possibility that $T \in [t, t + dt)$. If $T \geq t$, $Y(t) = 1$ and Equation 8.7.1 boils down to

$$\mathrm{pr}\{T < t + dt \mid T \geq t\} = h(t)dt,$$

a familiar equation. In this context, $\{Y(t)h(t)\}$ is the **intensity process** of N. The notation for the indicator process $Y(t)$ has become standard in this context, though a little unfortunate for statisticians who tend to think of Y as a response variable in a regression model. Its function in Equation 8.7.1 is to switch off the hazard once time $t = T$ is passed: after death, the intensity becomes zero, a fact of little comfort.

The compensator of the counting process N is the integrated intensity process $\Lambda(t)$, given by

$$\Lambda(t) = \int_0^t Y(s)h(s)ds. \tag{8.7.3}$$

Note that the processes Y, h, and Λ are all predictable with respect to $\{\mathcal{H}_t\}$. As shown in Section 8.5, $M(t) = N(t) - \Lambda(t)$ is a martingale counting process with respect to $\{\mathcal{H}_t\}$.

Suppose that the hazard function h depends on a parameter (vector) θ, so that

$$\Lambda(t;\theta) = \int_0^t Y(s)h(s;\theta)ds .$$

The likelihood function for θ based on observation of N is then constructed as follows. Either $dN(t) = 1$, indicating that failure is observed at time t, or $dN(t) = 0$, if not. The corresponding conditional probabilities, given the immediately preceding history, are

$$\text{pr}\{dN(t) = 1 \mid \mathcal{H}_{t-}\} = \text{E}\{dN(t) \mid \mathcal{H}_{t-}\} = d\Lambda(t;\theta) = Y(t)h(t;\theta)dt$$

and

$$\text{pr}\{dN(t) = 0 \mid \mathcal{H}_{t-}\} = 1 - d\Lambda(t;\theta).$$

The conditional likelihood for $dN(t)$ can then be written succinctly, in the usual fashion for a binary variate, as

$$d\Lambda(t;\theta)^{dN(t)}\{1 - d\Lambda(t;\theta)\}^{1-dN(t)} \propto \{Y(t)h(t;\theta)\}^{dN(t)}\{1 - Y(t)h(t;\theta)dt\}^{1-dN(t)},$$

where the dt-factor has been dropped from the first term. The overall likelihood function is now constructed as the product integral of these instantaneous conditional likelihoods over time. We will assume for now that the observation period is potentially infinite, without censoring, so that the actual failure time T is observed. Then the likelihood is

$$\begin{aligned} L(\theta) &= \mathcal{P}_{t\geq 0}[\{d\Lambda(t;\theta)\}^{dN(t)}\{1 - d\Lambda(t;\theta)\}^{1-dN(t)}] \\ &\propto \mathcal{P}_{t\geq 0}[\{Y(t)h(t;\theta)\}^{dN(t)}\{1 - Y(t)h(t;\theta)dt\}^{1-dN(t)}]. \end{aligned} \tag{8.7.4}$$

Now, $dN(t) = 0$ for all t except $t = T$, the observed failure time, when $dN(T) = 1$ and $Y(T) = 1$. Thus, the only factor not equal to 1 among the $\{Y(t)h(t;\theta)\}^{dN(t)}$ in Equation 8.7.4 is $h(T;\theta)$. The product integral over the other factors yields $\exp\{-\int_{t\geq 0} Y(t)h(t;\theta)dt\}$. Hence,

$$L(\theta) \propto h(T;\theta)\exp\left\{-\int_0^T h(t;\theta)dt\right\} = h(T;\theta)F(T;\theta) = f(T;\theta),$$

so $L(\theta)$ reduces to the familiar form in this most basic case of a single observed failure time.

8.7.2 *Independent lifetimes*

Consider now the case of n independent, not necessarily identically distributed, failure times, $T_1, \ldots, T_n$. Define counting processes $N_i(t) = \text{I}\{T_i \leq t\}$

and indicator processes $Y_i(t) = \mathrm{I}\{T_i \geq t\}$ for $i = 1, \ldots, n$. Also, define the multivariate counting process $N(t) = \{N_1(t), \ldots, N_n(t)\}$ and the associated filtration given by $\mathcal{H}_t = \sigma\{\bigcup_{i=1}^n \mathcal{H}_{it}\}$, where $\mathcal{H}_{it} = \sigma\{N_i(s): 0 \leq s \leq t\}$. Then N_i has compensator

$$\Lambda_i(t;\theta) = \int_0^t Y_i(s)h_i(s;\theta)ds$$

with respect to $\{\mathcal{H}_t\}$. The conditional probability that N_i jumps from 0 to 1 at time t, given the immediately preceding history, is

$$\mathrm{pr}\{dN_i(t) = 1 \mid \mathcal{H}_{t-}\} = \mathrm{E}\{dN_i(t) \mid \mathcal{H}_{t-}\} = d\Lambda_i(t;\theta) = Y_i(t)h_i(t;\theta)dt;$$

the conditional probability of no jump among the components of N at time t is

$$\mathrm{pr}\{dN_+(t) = 0 \mid \mathcal{H}_{t-}\} = 1 - \mathrm{E}\{dN_+(t) \mid \mathcal{H}_{t-}\} = 1 - d\Lambda_+(t;\theta),$$

where $dN_+(t) = \sum_{i=1}^n dN_i(t)$, etc.; in general, replacement of an index by + will be used to indicate summation over that index. Note the assumption implicit here that $dN_+(t)$ is 0 or 1, i.e., that there are no ties among the T_i. Thus, the conditional likelihood contribution for time t is

$$\Pi_{i=1}^n \{d\Lambda_i(t;\theta)\}^{dN_i(t)} \times \{1 - d\Lambda_+(t;\theta)\}^{1-dN_+(t)},$$

a multinomial expression analogous to the binomial one in Equation 8.7.4. The observation indicator $Y_i(t)$ is equal to 1 on $[0, T_i]$ and to 0 on (T_i, ∞). The likelihood function for observation of $N_1, \ldots, N_n$ is then obtained as the product integral

$$L(\theta) = \mathcal{P}_{t \geq 0}\left[\prod_{i=1}^n \{d\Lambda_i(t;\theta)\}^{dN_i(t)} \times \{1 - d\Lambda_+(t;\theta)\}^{1-dN_+(t)}\right]$$

$$\propto \mathcal{P}_{t \geq 0}\left[\prod_{i=1}^n \{Y_i(t)h_i(t;\theta)\}^{dN_i(t)} \times \{1 - \sum_{i=1}^n Y_i(t)h_i(t;\theta)dt\}^{1-dN_+(t)}\right].$$

Now, $dN_i(t) = 0$ for all t except $t = T_i$, the observed failure time, when $dN_i(T_i) = 1$ and $Y_i(T_i) = 1$. So, $\mathcal{P}_{t \geq 0}\{Y_i(t)h_i(t;\theta)\}^{dN_i(t)}$ reduces to $h_i(T_i;\theta)$. Again, $dN_+(t)$ is equal to 0 for all $t \geq 0$ except at the times $t = T_i$. Thus, the product integral over the other term in $L(\theta)$ yields $\exp\{-\int_{t \geq 0} \sum_{i=1}^n Y_i(t)h_i(t;\theta)dt\}$. Hence,

$$L(\theta) \propto \prod_{i=1}^n \{h_i(T_i;\theta)\} \times \exp\left\{-\sum_{i=1}^n \int_0^{T_i} h_i(t;\theta)dt\right\}$$
$$= \prod_{i=1}^n \{h_i(T_i;\theta)\bar{F}_i(T_i;\theta)\} = \Pi_{i=1}^n f_i(T_i;\theta). \tag{8.7.5}$$

In the independent, identically distributed case, where $h_i(t;\theta) = h(t;\theta)$ for all i, $N_+(t)$ is a counting process with compensator

$$\Lambda_+(t;\theta) = \int_0^t Y_+(s)h(s;\theta)ds,$$

where $Y_+(t)$ is the number of T_is greater than or equal to t. Thus, $Y_+(t)$ is the number of individuals "at risk" at time $t-$; earlier, you will recall, this was denoted by q, e.g., Section 4.3.1.

8.7.3 Competing risks

In order to deal with p competing risks, we set up a p-variate counting process for each individual: for $i = 1, \ldots, n$, set

$$N_i(t) = \{N_{i1}(t), \ldots, N_{ip}(t)\},$$

where $N_{ij}(t)$ corresponds to the jth risk. Thus, at the failure time $t = T_i$, when (C_i, T_i) is observed, the C_ith component of $N_i(t)$ jumps from 0 to 1, and the others remain at 0. An overall multivariate counting process N is now defined having ith component vector $N_i(t)$. The framework can then be described in terms of filtrations $\mathcal{H}_{it} = \sigma\{N_i(s): 0 \le s \le t\}$ and an overal filtration $\mathcal{H}_t$ containing them. The corresponding compensators are, for $i = 1, \ldots, n$ and $j = 1, \ldots, p$,

$$\Lambda_i(j, t;\theta) = \int_0^t Y_i(s)h_i(j, s;\theta)ds,$$

where the $h_i(j, s;\theta)$ are the sub-hazards for individual i. The conditional probability that N_{ij} jumps from 0 to 1 at time t, given the immediately preceding history, is

$$\text{pr}\{dN_{ij}(t) = 1 \mid \mathcal{H}_{t-}\} = \text{E}\{dN_{ij}(t) \mid \mathcal{H}_{t-}\} = d\Lambda_i(j, t;\theta) = Y_i(t)h_i(j, t;\theta)dt;$$

the conditional probability of no jump among the np components of N at time t is

$$\text{pr}\{dN_{++}(t) = 0 \mid \mathcal{H}_{t-}\} = 1 - \text{E}\{dN_{++}(t) \mid \mathcal{H}_{t-}\} = 1 - d\Lambda_+(+, t;\theta).$$

Thus, the conditional likelihood contribution for time t is the multinomial expression

$$\Pi_{i=1}^n \Pi_{j=1}^p \{d\Lambda_i(j, t;\theta)\}^{dN_{ij}(t)} \times \{1 - d\Lambda_+(+, t;\theta)\}^{1 - dN_{++}(t)}.$$

The overall likelihood function is then

$$L(\theta) = \mathcal{P}_{t\geq 0}\left[\prod_{i=1}^{n}\prod_{j=1}^{p}\{d\Lambda_i(j,t;\theta)\}^{dN_{ij}(t)} \times \{1 - d\Lambda_+(+,t;\theta)\}^{1-dN_{++}(t)}\right]$$

$$\propto \mathcal{P}_{t\geq 0}\left[\prod_{i=1}^{n}\prod_{j=1}^{p}\{Y_i(t)h_i(j,t;\theta)\}^{dN_{ij}(t)} \times \left\{1 - \sum_{i=1}^{n}\sum_{j=1}^{p} Y_i(t)h_i(j,t;\theta)dt\right\}^{1-dN_{++}(t)}\right]$$

In the case of pure competing risks with no ties, only one component of N_i can jump at any one time, and when this happens for the first time observation ceases on individual i. If the outcome for the ith individual is (C_i, T_i), the only non-zero value among the $dN_{ij}(t)$ (for $j = 1, \ldots, p$) is that for $j = C_i$ and $t = T_i$, when $dN_{ij}(T_i) = 1$ and $Y_i(T_i) = 1$. Thus, the term $\mathcal{P}_{t\geq 0}\,\Pi_{j=1}^{p}\{Y_i(t)h_i(j,t;\theta)\}^{dN_{ij}(t)}$ in $L(\theta)$ reduces to $h_i(C_i, T_i;\theta)$. In the other term, $dN_{++}(t)$ is also equal to 0 for all $t > 0$ except at each T_i, when it takes value 1; we assume that $dN_{++}(t)$ cannot be greater than 1, i.e., that there are no ties among the T_i. Also, $Y_i(t)$ is equal to 1 on $[0, T_i]$ and to 0 on (T_i, ∞). Thus,

$$\begin{aligned} L(\theta) &= \prod_{i=1}^{n}\{h_i(C_i,T_i;\theta)\} \times \exp\left\{-\sum_{i=1}^{n}\sum_{j=1}^{p}\int_{t\geq 0} Y_i(t)h_i(j,t;\theta)dt\right\} \\ &= \prod_{i=1}^{n}\{h_i(C_i,T_i;\theta)\} \times \exp\left\{-\sum_{i=1}^{n}\int_0^{T_i} h_i(t;\theta)dt\right\} \qquad (8.7.6) \\ &= \prod_{i=1}^{n}\{h_i(C_i,T_i;\theta)\bar{F}_i(T_i;\theta)\} = \Pi_{i=1}^{n} f_i(C_i,T_i;\theta). \end{aligned}$$

8.7.4 Right-censoring

Some notes were made in Section 7.6, essentially setting the censoring mechanism in context as an additional competing risk. The known identifiability results for competing risks can then be applied to determine whether censoring obstructs inference about the target failure time distribution. The simplest cases yielding non-obstruction are where the censoring time S is fixed in advance, such as in a study of pre-determined time period, or where it is generated by some stochastic mechanism that runs independently of the one generating the failure time T. Even in this second situation, however, censoring might be **informative** in the sense that the S-distribution could depend on a parameter that also appears in the T-distribution. In this case, ignoring the S-contribution produces a partial likelihood.

The explicit effect of censoring on the likelihood Equation 8.7.6 is as follows. In the penultimate expression, $\Pi_{i=1}^{n}\{h_i(C_i, T_i;\theta)\bar{F}_i(T_i;\theta)\}$, the factor $h_i(C_i, T_i;\theta)$ is just equal to 1 at a pre-determined censoring time, with $C_i = 0$ in our notation. Then, the likelihood contribution from the ith individual is, familiarly, $\bar{F}_i(T_i;\theta)$ instead of $f_i(T_i;\theta)$. In the case of stochastic censoring, this could have been included as an additional competing risk by extending the factor in the original expression for $L(\theta)$ to $\Pi_{i=1}^{n}\Pi_{j=0}^{p}\{d\Lambda_i(j,t;\theta)\}^{dN_{ij}(t)}$, including a factor for risk 0. If the ith individual failure time is censored, the resulting contribution is $h_i(0, T_i;\theta)$. The omission of such factors from $L(\theta)$ is effectively the same as setting them equal to 1, and this yields $\bar{F}_i(T_i;\theta)$ in place of $f_i(T_i;\theta)$ in the final expression, as before. If the censoring hazard $h_i(0, t;\theta)$ does not depend on θ, nothing is lost in $L(\theta)$ by omitting it; if it does, its omission makes $L(\theta)$ a partial likelihood.

8.8 Non-parametric estimation

8.8.1 Survival times

Consider a random sample of failure times, ordered as $T_1 < T_2 < \ldots < T_n$. Let $N_+(t)$ be the associated counting process, taking value i on $[T_i, T_{i+1})$ for $i = 0, 1, \ldots, n$; by convention, $T_0 = 0$ and $T_{n+1} = \infty$. The predictable process $Y_+(t)$ indicates the number of individuals "at risk" at time $t-$: $Y_+(t) = n - i$ on $(T_i, T_{i+1}]$ for $i = 0, 1, \ldots, n$, assuming that there is no censoring. Let $M_+(t) = N_+(t) - \Lambda_+(t)$ be the corresponding martingale counting process. Then, from $dN_+(t) = d\Lambda_+(t) + dM_+(t)$, the natural (conditional moment) estimator of $d\Lambda_+(t)$ is $dN_+(t)$, since $\mathrm{E}\{dM_+(t) \mid \mathcal{H}_{t-}\} = 0$ confers upon $dM_+(t)$ the status of a zero-mean "error term." But, $d\Lambda_+(t) = Y_+(t)h(t)dt$, so the corresponding estimator for the integrated hazard function $H(t) = \int_0^t h(s)ds$ is $H(t) = \int_0^t Y_+(s)^{-1} dN_+(s)$. To avoid the problem caused here when $Y_+(s) = 0$, the **Nelson-Aalen estimator** is defined as

$$\tilde{H}(t) = \int_0^t Y_+(s)^- dN_+(s); \qquad (8.8.1)$$

here, $Y_+(s)^- = Y_+(s)^{-1}$ when $Y_+(s) > 0$ and $Y_+(s)^- = 0$ when $Y_+(s) = 0$, using a notation analogous to that for a generalised inverse matrix. This integral actually reduces to a sum:

$$\tilde{H}(t) = \sum\nolimits_t Y_+(T_i)^-,$$

where $\sum_t$ denotes summation over $\{i{:}T_i \le t\}$. Thus, $\tilde{H}(t)$ is a step function: as t increases, $\tilde{H}(t)$ steps up by $Y_+(T_i)^-$ as each T_i is reached: explicitly, on $[T_i, T_{i+1})$, $\tilde{H}(t)$ takes value

$$Y_+(T_1)^- + \ldots + Y_+(T_i)^- = n^{-1} + (n-1)^{-1} + \ldots + (n-i+1)^{-1}$$

for $i = 1, \ldots, n$; on $[0, T_1)$, $\tilde{H}(t) = 0$.

We have

$$\tilde{H}(t) = \int_0^t Y_+(s)^- dN_+(s) = \int_0^t Y_+(s)^- \{d\Lambda_+(s) + dM_+(s)\}$$

$$= \int_0^t Y_+(s)^- Y_+(s)h(s)ds + \int_0^t Y_+(s)^- dM_+(s).$$

Thus,

$$\tilde{H}(t) - H(t) = \int_0^t \{Y_+(s)^- Y_+(s) - 1\}h(s)ds + \int_0^t Y_+(s)^- dM_+(s). \quad (8.8.2)$$

But, $Y_+(s)^- Y_+(s)$ is equal to 1 if $Y_+(s) > 0$, and to 0 if $Y_+(s) = 0$. Therefore,

$$E\{Y_+(s)^- Y_+(s)\} = pr\{Y_+(s) > 0\} = 1 - pr\{Y_+(s) = 0\},$$

and so Equation 8.8.2 yields

$$E\{\tilde{H}(t)\} = H(t) - \int_0^t pr\{Y_+(s) = 0\}h(s)ds,$$

the expectation of the second term, a stochastic integral with respect to the martingale M_+, being zero. Thus, $\tilde{H}(t)$ is approximately unbiased for $H(t)$ if $pr\{Y_+(s) = 0\}$ is small on $[0, t)$, and this will be true when $pr(T_n \le t)$ is small. In fact, $pr(T_n \le t) \to 0$ as $n \to \infty$ for fixed t when the failure time variate has positive probability beyond t.

From Equation 8.4.8, the predictable variation process for $\tilde{H}(t)$ can be approximated as $\int_0^t \{Y_+(s)^-\}^2 d\langle M_+(s)\rangle$. We now have approximate expressions for the mean and variation of $\tilde{H}(t)$, and these can be employed, in conjunction with a central limit theorem for martingales (e.g., Andersen et al., 1993, Section IV.1.2) to obtain tests and confidence intervals via the usual asymptotic ($n \to \infty$) approximations.

An estimator for the survivor function

$$\bar{F}(t) = \exp\left\{-\int_0^t h(s)ds\right\} = \mathcal{P}_{s \le t}\{1 - dH(s)\}$$

can now be obtained by replacing $dH(s)$ here by $d\tilde{H}(s) = Y_+(s)^- dN_+(s)$. Thus,

$$\tilde{\bar{F}}(t) = \mathcal{P}_{s \le t}\{1 - Y_+(s)^- dN_+(s)\} = \Pi_t\{1 - Y_+(T_i)^{-1}\},$$

where Π_t is the product over $\{i: T_i \le t\}$. This is just the **Kaplan-Meier estimator** (Section 4.3.1).

8.8.2 *Competing risks*

The extension of the above to the case of p competing risks is effected by defining a p-variate counting process $(N_{+1}, \ldots, N_{+p})$, as in Section 8.7.3. We define $H(j, t) = \int_0^t h(j, s)ds$, the integrated sub-hazard function, and the Nelson-Aalen-type estimator

$$\tilde{H}(j, t) = \int_0^t Y_+(s)^- dN_{+j}(s).$$

Now, N_{+j} jumps at the tmes of j-type failures, say T_{1j}, T_{2j}, ..., and then $H(j, t)$ reduces to $\sum_t Y_+(T_{ij})^-$, where the summation is over $\{i\colon T_{ij} \le t\}$. Explicitly,

$$\tilde{H}(j, t) = q(T_{1j})^{-1} + \ldots + q(T_{i_t j})^{-1},$$

where $q(t)$ is the size of the risk set at time t and $i_t = \max\{i\colon T_{ij} \le t\}$; compare this with Equation 6.3.3.

We can now adopt $M_{+j}(t) = N_{+j}(t) - \Lambda_+(j, t)$ and

$$\tilde{H}(j, t) - H(j, t) = \int_0^t \{Y_+(s)^- Y_+(s) - 1\} h(j, s) ds + \int_0^t Y_+(s)^- dM_{+j}(s)$$

in order to develop methods analogous to those outlined in the preceding section.

8.9 Non-parametric testing

Let T_{ij} be the failure or censoring time and C_{ij} be the censoring indicator (1 if failure observed, 0 if right-censored) for individual i ($i = 1, \ldots, n_j$) in group j ($j = 1, \ldots, g$). We assume that the individual (T_{ij}, C_{ij}) are all independent and that the censoring is independent. Let $N_{ij}(t)$ be the individual counting processes and $Y_{ij}(t) = \mathrm{I}\{T_{ij} \ge t\}$ be the associated indicator processes, and let $h_j(t)$ be the hazard function for individuals in group j. Define

$$N_{+j}(t) = \sum_{i=1}^{n} N_{ij}(t), \quad d\Lambda_{+j}(t) = \sum_{i=1}^{n} Y_{ij}(t) h_j(t) = Y_{+j}(t) h_j(t),$$

and

$$M_{ij}(t) = N_{ij}(t) - \Lambda_{ij}(t),$$

the $M_{ij}(t)$ being martingales with respect to the filtration $\sigma\{\bigcup_{ij} \mathcal{H}_{ij}\}$. The Nelson-Aalen estimator for the integrated hazard function in the jth group is

$$\tilde{H}_j(t) = \int_0^t Y_{+j}(s)^- dN_{+j}(s).$$

Under H_0: $h_j = h$ ($j = 1, \ldots, g$), a pooled estimator for the common integrated hazard function is

$$\tilde{H}(t) = \int_0^t Y_{++}(s)^- dN_{++}(s).$$

Define

$$H_j^*(t) = \int_0^t \mathrm{I}\{Y_{+j}(s) > 0\} Y_{++}(s)^- dN_{++}(s),$$

then

$$\begin{aligned}\tilde{H}_j(t) - H_j^*(t) &= \int_0^t Y_{+j}(s)^- \{dM_{+j}(s) + d\Lambda_{+j}(s)\} \\ &\quad - \int_0^t \mathrm{I}\{Y_{+j}(s) > 0\} Y_{++}(s)^- \{dM_{++}(s) + d\Lambda_{++}(s)\} \\ &= \int_0^t Y_{+j}(s)^- dM_{+j}(s) - \int_0^t \mathrm{I}\{Y_{+j}(s) > 0\} Y_{++}(s)^- dM_{++}(s) \\ &\quad + \int_0^t [Y_{+j}(s)^- d\Lambda_{+j}(s) - \mathrm{I}\{Y_{+j}(s) > 0\} Y_{++}(s)^- d\Lambda_{++}(s)].\end{aligned}$$

Note that, under H_0,

$$Y_{+j}(s)^- d\Lambda_{+j}(s) = Y_{+j}(s)^- Y_{+j}(s) h(s) ds = \mathrm{I}\{Y_{+j}(s) > 0\} h(s) ds,$$

and

$$\mathrm{I}\{Y_{+j}(s) > 0\} Y_{++}(s)^- d\Lambda_{++}(s) = \mathrm{I}\{Y_{+j}(s) > 0\} Y_{++}(s)^- Y_{++}(s) h(s) ds = \mathrm{I}\{Y_{+j}(s) > 0\} h(s) ds.$$

Hence, under H_0, the third integral on the right hand side is zero, leaving $H_j(t) - H_j^*(t)$ as the difference of two stochastic integrals with respect to martingales, i.e., a martingale itself. Now define a further martingale,

$$Z_j(t) = \int_0^t K_j(s) \{d\tilde{H}_j(t) - dH_j^*(t)\}, \tag{8.9.1}$$

by taking $K_j(s)$ as some predictable process; K_j is interpreted as a weighting function applied to the differences between $\tilde{H}_j$ and H_j^*, and $Z_j(t)$ as a weighted sum (integral) of these differences. Different choices for K_j, together with subsequent combination of the Z_j, produce different test statistics for H_0.

One choice for $K_j(s)$ is $Y_{+j}(s)\mathrm{I}\{Y_{++}(s) > 0\}$. Then,

$$\begin{aligned}Z_j(t) &= \int_0^t Y_{+j}(s) \mathrm{I}\{Y_{++}(s) > 0\} \\ &\quad [Y_{+j}(s)^- dN_{+j}(s) - \mathrm{I}\{Y_{+j}(s) > 0\} Y_{++}(s)^- dN_{++}(s)] \qquad (8.9.2) \\ &= \int_0^t Y_{+j}(s)^- Y_{+j}(s) dN_{+j}(s) - \int_0^t Y_{++}(s)^- Y_{+j}(s) dN_{++}(s).\end{aligned}$$

The first term on the right hand side is the observed number of failures in group j during $[0, t]$, and the second is an estimate of its expected value under H_0. Setting $t = \infty$, and combining the $Z_j(\infty)$ into a chi-square statistic of familiar form $z^T V_z^{-1} z$, produces the famous **log-rank test** (Peto and Peto, 1972). Different choices for K_j in Equation 8.9.1 give rise to a class of such tests, the members of which will have different detailed properties.

8.10 Regression models

8.10.1 Intensity models and time-dependent covariates

The Cox proportional hazards model for competing risks (Section 6.4.1) specifies the sub-hazard function $h(j, t;x)$ to be of the form $\psi_j(x;\beta_j)h_0(j, t)$, in which $\psi_j(x;\beta_j)$ is a positive function of the covariate x and parameter β_j, often taken to be $\exp(x^T\beta_j)$, and $h_0(j, t)$ is a baseline sub-hazard. The β_j $(j = 1, \ldots, p)$ are sub-vectors, not necessarily disjoint, of an overall regression vector β. In the present context the model would be written as

$$d\Lambda(j, t;\beta_j) = Y(t)\ \psi_j(x;\beta_j)\ \lambda_0(j, t)\ dt.$$

This is a type of **multiplicative intensity model**. Had we begun with the form $\psi_j(x;\beta_j) + h_0(j, t)$ for the sub-hazard, an **additive intensity model** would have resulted. As usual in regression, the inferences will normally be made conditionally on the x-values in the sample. Otherwise, their distribution has to be accommodated in the likelihood function, which might or might not be a simple matter.

The covariate vector is assumed to be constant over time in the above, but this is not necessary and is often violated in practice. Time-dependent covariates were discussed in Section 4.2. The form is $x(t)$, with sample path over time governed by some stochastic mechanism that may be connected with that generating the failure time itself. Broadly speaking, if $x(t)$ is determined by the past history, i.e., is measurable with respect to $\mathcal{H}_{t-}$, it can be included in the likelihood without too much trouble: its predictability, like that of $Y(t)$, allows one to insert its value into the likelihood contribution at time t. Note that the filtration $\{\mathcal{H}_t\}$ might have to be augmented to accommodate the x-process.

Oakes (1981) and Arjas (1985) have made important points concerning the definition of "time" in Survival Analysis, one of which can be illustrated as follows. Suppose that certain time-varying conditions affect all individuals in a study, environmental factors being a case in point. Further, suppose that different individuals enter at different dates (staggered entry) and that a proportional hazards model with unspecified baseline hazard $h_0(t)$ is adopted that is intended to account for some of the unrecorded time-varying conditions. Then, the usual time-shift, making each individual start at $t = 0$, disrupts the synchrony in $h_0(t)$ between them. In effect, times in $h_0(t)$ and the $x_i(t)$ are different.

8.10.2 Proportional hazards model

The Cox model for competing risks is defined by

$$d\Lambda_i(j, t) = Y_i(t)\, h_i(j, t)\, dt = Y_i(t)\, \psi_j(x_i(t);\beta_j)\, h_0(j, t;\phi)\, dt,$$

where the covariate vector x may vary with t and the baseline hazard h_0 may depend on a parameter vector ϕ. To construct a likelihood function $L(\theta)$, where $\theta = (\beta, \phi)$, we set up a multivariate counting process $N = (N_1, \ldots, N_n)$, with $N_i = (N_{i1}, \ldots, N_{ip})$, as in Section 8.7.3. From the first expression in Equation 8.7.6, the log-likelihood can be written down as

$$\begin{aligned}\log L(\theta) &= \sum_{i=1}^{n} \log h_i(c_i, t_i) - \sum_{i=1}^{n}\sum_{j=1}^{p}\int_{t\geq 0} Y_i(t) h_i(j, t)dt \\ &= \sum_{i=1}^{n}\sum_{j=1}^{p}\left[\int_{t\geq 0} Y_i(s)\log\{h_i(j, s)\}dN_{ij}(s) - \int_{t\geq 0} Y_i(s) h_j(j, s)ds\right].\end{aligned}$$

The theoretical manipulations now follow from seeing that the score function, $d\log L(\theta)/d\theta$, has components that are linear combinations of stochastic integrals with respect to the martingales $M_{ij}(t) = N_{ij}(t) - \Lambda_i(j, t)$ and are therefore themselves martingales.

To tackle the semi-parametric version, in which the $h_0(j, t;\phi)$ are treated as unspecified nuisance functions, we adopt Cox's partial likelihood function, $P(\beta)$ in Equation 6.4.2, as the basis for inference. It is then found that the partial score vector $dP/d\beta$ is a martingale and the rest, as they say, is geography.

8.10.3 Martingale residuals

Various types of residuals can be defined for assessing model fit, both overall and for individual cases. In addition to those discussed in Section 2.2.2, residuals associated with the counting process approach are described and exploited by Fleming and Harrington (1991, Section 4.5).

In a random sample of failure times (Section 8.7.2), the process $M_i(t) = N_i(t) - \Lambda_i(t;\theta)$ is a martingale specific to individual i. At the end of the study, at time τ, say, $N_i(\tau)$ will be 1 if failure has been observed and 0 if not. The differences, $N_i(\tau) - \Lambda_i(\tau;\hat{\theta})$, are known as **martingale residuals**: they can be plotted and tested in various ways.

The possible values assumed by $N_i(t)$ in standard Survival Analysis are 0 and 1, so the range of values for $M_i(t)$ is $(-\infty, 1)$ and such martingale residuals can tend to have negatively skewed sampling distributions. Some calculations, along the lines that produce **deviance residuals** in the context of generalised linear models, produce a transformed version of $M_i(t)$ that has a more symmetric distribution. The transformation (Fleming and Harrington, 1991, Equation 5.14), has the effect of shrinking large negative values towards 0 and expanding values near 1 away from 0. Other types described in that book include **martingale transform residuals**, of which a particular species is **score residuals**, and the authors show how to use such residuals to assess influence of individual cases and model adequacy. Additional material on residuals is to be found in Chapter 7 of Andersen et al. (1993). **Schoenfeld residuals** are particularly aimed at detecting non-proportionality in the proportional hazards model (Section 8.10.2). In Schoenfeld (1982) and Grambsch and Therneau (1994), they are used for plotting and the construction of test statistics.

8.11 *Epilogue*

What are the directions of development for the subject? I'm not normally one to be drawn into grandiose speculation — it reminds me of Peter Sellers' spoof political speech "The Future Lies Ahead" — come to think of it, it reminds me of most political speeches. However, it has been suggested that there should be a few words on this at the end of the book, so here goes. There have been some obviously identifiable key advances, such as non-parametric maximum likelihood (Kaplan and Meier, 1958), semi-parametric modelling (Cox, 1972), and martingale counting processes (Aalen, 1975). The strength of such works is demonstrated by the huge research literature that they have spawned. Also, it is in the sequential nature of research that there is a wealth of groundwork to be drawn on. For instance, some of what we now call non-parametric methods have been used from time immemorial in the actuarial business. So, what are the star turns of the future? One contender is the use of much more sophisticated non-parametric and semi-parametric methodology. For example, Walker et al. (1999) presented a modelling framework for the integrated hazard function based on a particular kind of stochastic process. The aim is to have a flexible prior distribution on the space of distribution functions that gives rise to tractable (conjugate) posterior forms in a Bayesian framework. The approach is mathematically advanced, but, then, who would have thought that martingales and counting processes would ever become standard tools in applied statistics? In a slightly different direction, there is the semi-parametric approach of Couper and Pepe (1997), in which the modern technique of Generalised Additive Models is applied.

On a more personal level, if you have stayed the course, battled through thick and thin to this point, may I congratulate you. I hope that you have come to appreciate the subject of Competing Risks and wish to go further and deeper into it. The immediate way ahead, I suggest, is to continue from where the present chapter leaves off, and a good way of doing this is to work through one or both of Fleming and Harrington (1991) and Andersen et al. (1993). The counting process approach has solid advantages: it provides a firm theoretical basis for modelling events unfolding over time and is particularly suited to models framed in terms of hazard rates and involving censoring. I can do no better than quote from the Preface of Fleming and Harrington (1991): "Martingale methods can be used to obtain simple expressions for the moments of complicated statistics, to calculate and verify asymptotic distributions for test statistics and estimators, to examine the operating characteristics of non-parametric testing methods and semi-parametric censored data regression methods, and even to provide a basis for graphical diagnostics in model building with counting process data." Good studying!

appendix 1

Numerical maximisation of likelihood functions

Suppose that we have a data set comprising n observations, and a model involving a parameter vector θ. Denote the likelihood function by $L_n(\theta)$. Maximum likelihood estimates are found by maximising $L_n(\theta)$ over θ. In "regular" cases, this can be done by differentiating $\log L_n$ with respect to θ to obtain the **score function**, a vector of length $\dim(\theta)$, and then the **likelihood equations** result from setting the score function equal to the zero vector. Often, these equations can be solved explicitly for at least some θ-components, thus expressing a subset of the maximum likelihood estimates in terms of the rest and thus reducing the dimension of the maximisation problem.

For complicated likelihood functions, it is often more convenient to apply a function optimization routine, such as subroutine e04jaf of the NAG (1988) library, directly, rather than differentiating analytically to obtain the likelihood equations and then solving them numerically. The direct scheme is obviously less trouble to implement and less error prone. It provides fewer of those golden opportunities for algebraic and program-coding slips that we all so eagerly grasp. It is also possibly more accurate numerically, provided that the subroutine to compute $L_n(\theta)$ on call has been carefully coded to minimize cancellation errors and the like. This is because gradient-based algorithms such as e04jaf compute derivatives by finite differencing, the operation being controlled to give values that might well be more accurate than one's own coding.

In using general optimization algorithms, it is always worthwhile to take some care with the parametrization. For instance, scale the parameters so that the effect on the log-likelihood of changing any of their values by, say, 0.1 is about the same. Even more importantly, give them some elbow room, e.g., if a parameter α is constrained to lie between 0 and 1, re-parametrize in terms of $\alpha' = \log\{\alpha/(1-\alpha)\}$ which is now free to roam the wide-open spaces. Ross (1970) has some useful advice on this topic.

To give a brief outline of the direct-optimization approach, suppose that a general, twice-differentiable function $G(\theta)$ is to be minimized over θ; for maximum likelihood, take $G(\theta) = -\log L_n(\theta)$. We begin with an initial estimate θ_1 of the minimum point and then update it to $\theta_2 = \theta_1 + s_1 \boldsymbol{d}_1$, where $\boldsymbol{d}_1$ is the step direction vector and s_1 is the step length. The procedure is iterated, taking θ_2 as the new initial estimate and so on. A sequence $\{\theta_1, \theta_2, \theta_3, \ldots\}$ is thus generated which, we hope, will converge rapidly to the minimum point sought. Termination of the search will be suggested when the sequences $\{\theta_j\}$ and $\{G(\theta_j)\}$ have stopped moving significantly, and g_j, denoting the gradient vector $\boldsymbol{G}' = dG/d\theta$ evaluated at θ_j, is near zero.

Since we wish to move downhill, thinking of $G(\theta)$ as a surface above the θ-plane, $\boldsymbol{d}_j$ should be closely related to $-g_j$. Commonly, $\boldsymbol{d}_j$ is taken as $-\boldsymbol{A}_j g_j$, where $\boldsymbol{A}_j$ is a matrix used to modify the step direction in some advantageous way. By Taylor expansion, and using $\theta_{j+1} = \theta_j - s_j \boldsymbol{A}_j g_j$,

$$G(\theta_{j+1}) \approx G(\theta_j) + \boldsymbol{G}'(\theta_j)^{\mathrm{T}}(\theta_{j+1} - \theta_j) = G(\theta_j) - s_j g_j^{\mathrm{T}} \boldsymbol{A}_j g_j,$$

so that $G(\theta_{j+1}) < G(\theta_j)$ if $\boldsymbol{A}_j$ is positive definite and $g_j \neq 0$. The scaling factor s_j is used to adjust the step length to locate a minimum, or just to achieve a reduction in $G(\theta)$, along the step direction; it is determined by a linear search in this direction. If $\boldsymbol{A}_j$ is positive definite, then a reduction is ensured for small enough s_j, but a larger value of s_j might give a larger reduction and lead to quicker minimization. There are various standard choices for $\boldsymbol{A}_j$.

(i) $\boldsymbol{A}_j = \mathbf{I}$ (unit matrix) yields the method of **steepest descents**. Unfortunately, over successive steps the search path can come to resemble hem-stitching, i.e., a zig-zag, crossing and re-crossing the valley floor, approaching the minimum point painfully slowly. The problem here is one of retracing one's steps, i.e., the search direction's not breaking fresh ground, not moving into subspaces unspanned by previous search directions.

(ii) $\boldsymbol{A}_j = (\boldsymbol{G}_j'')^{-1}$, where $\boldsymbol{G}_j'' = \boldsymbol{G}''(\theta_j)$, yields **Newton's**, or the **Newton-Raphson**, method. This arises from the observation that, if θ_{j+1} is actually the minimum point, then

$$\mathbf{0} = \boldsymbol{G}'(\theta_{j+1}) \approx \boldsymbol{G}'(\theta_j) + \boldsymbol{G}_j''(\theta_{j+1} - \theta_j),$$

by Taylor expansion. Hence,

$$\theta_{j+1} \approx \theta_j - (\boldsymbol{G}_j'')^{-1} \boldsymbol{G}'(\theta_j) = \theta_j - \boldsymbol{A}_j g_j.$$

If $\boldsymbol{G}(\theta)$ is a positive definite quadratic function of θ, $\boldsymbol{G}''(\theta)$ is a constant matrix, the Taylor expansion is exact, and the minimum will be reached in one step from any starting point. However, for a non-quadratic function, if $\boldsymbol{G}_j''$ is not positive definite at θ_j, this search direction might be totally useless. Even if $\boldsymbol{G}_j''$ is positive definite, its computation and inversion at each step is costly.

(iii) In the celebrated **Levenburg-Marquardt compromise** (Levenburg, 1944; Marquardt, 1963), between Newton-Raphson and steepest descents, A_j is taken as $(G_j'' + \lambda_j I)^{-1}$, where λ_j is a tuning constant. For sufficiently large λ_j, $G_j'' + \lambda_j I$ will be positive definite. Statisticians have heard an echo of this in **Ridge Regression**.

(iv) Another variant is **Fisher's scoring method** in which $G(\theta)$ depends on sample values of random variables and A_j is taken as $\{E(G_j'')\}^{-1}$. In certain special situations, such as when $G(\theta)$ is the log-likelihood function for an exponential family model, this can yield a worthwhile simplification (McCullagh and Nelder, 1989, Section 2.5).

(v) **Conjugate direction** methods set out to cure the deficiency mentioned in (i) by making the search direction d_j orthogonal to the previous ones. For instance, we might modify the basic steepest descents process by taking $d_1 = -s_1 g_1$, then $d_2 = -s_2 g_2 + t_1 d_1$, with t_1 chosen to make $d_1^T d_2 = 0$, i.e., $t_1 = s_1 d_1^T g_2 / d_1^T d_1$, and so on. There are many cunning variants, each with its own scheme for generating the d_j, and each with its own strengths and weaknesses.

(vi) In taking $-A_j g_j$ for the step direction, rather than just $-g_j$, one is effectively altering the metric of the θ-space. For a positive definite quadratic function, the contours of constant G are elliptical, and taking $d_j \propto -(G_j'')^{-1} g_j$ effectively makes them circular, so that the steepest descent direction points straight at the minimum. Davidon (1959) coined the term "**variable metric**" for methods in which the "metric correcting transformation" A_j is updated at each step. As the neighbourhood of the minimum is approached, A_j should tend to $(G_j'')^{-1}$ since G will be approximately quadratic there, which is when Newton's method comes into its own. Fletcher and Powell (1963) developed Davidon's idea, and, with a particular updating scheme for A_j, the method found fame as the **DFP algorithm**. Much work has followed and there is now a vast literature on sophisticated **quasi-Newton methods**. The book by Gill, Murray, and Wright (1981) gives a pretty comprehensive survey up to 1980.

In the so-called orthogonal case (Cox and Reid, 1987), the log-likelihood breaks down into a sum of contributions involving separate subsets of parameters:

$$G(\theta) = -\log L_n(\theta) = G^{(1)}(\theta^{(1)}) + \ldots + G^{(r)}(0^{(r)}),$$

say. Then G_j'', or any of its variants here, will be block-diagonal. As a result, the r component steps in d_j will be computed independently. Thus, the minimisation will proceed almost as if the contributions were being dealt with separately: almost, because the determination of s_j, the step length adjustment, will introduce some degree of compromise. However, if storage is not a problem, minimisation of $G(\theta)$ in one go will be hardly any slower than giving each contribution special treatment.

There is an unspoken assumption in all of this that the value located by the optimisation algorithm is the maximum likelihood estimate. In practice, however, it is not uncommon for the likelihood function to have local maxima to which an algorithm can easily be drawn. Such problems can be addressed by more sophisticated methods such as simulated annealing (Brooks and Morgan, 1995), which are more likely to locate global maxima. However, the existence of multiple maxima can indicate that the asymptotic theory, on which the usefulness of maximum likelihood estimation relies, is not effective for the situation; for instance, the sample size might be too small or the likelihood might be non-regular. This is a real problem and there are no simple, general solutions. (In theory, it is not a problem for the Bayesian approach.)

To summarise, powerful methods for function optimisation have been developed over many years by the professionals, expert numerical analysts. Algorithms for conjugate directions and variable metric methods are implemented in computer packages such as the NAG library. It seems a little odd that, even nowadays, so many pages of statistical journals and books are given over to comparatively amateur accounts by statisticians for particular applications. An excellent general book on numerical methods, described in simple terms and giving actual computer routines to implement them, is that of Press et al. (1990).

appendix 2

Bayesian computation

The statistical literature is now rich in accounts of numerical methods for Bayesian computation, both at the introductory and research levels. This contrasts with the situation in function optimisation, which is mainly in the mathematical and operational research literature. Therefore, we give here only a very brief account.

As pointed out in Section 2.3.3, many required quantities are expressible as posterior expectations. Suppose that we have a random sample $(\theta_1,...,\theta_m)$ from the posterior distribution $p(\theta \mid D)$ of θ, D denoting the data on which it is based. Then, for any given function $\eta(\theta)$, $E\{\eta(\theta) \mid D\}$ can be estimated as $m^{-1}\sum_{j=1}^{m}\eta(\theta_j)$. The main thrust of the numerical methods is to produce such a random sample of θ-values or, at least, something sufficiently close to it.

In Importance Sampling and the Rejection Method, a random sample is drawn, not from $p(\theta \mid D)$ itself, but from some other convenient and suitable density $q(\theta)$. The values generated are then modified, effectively to bring the summation back to the required form. For Importance Sampling this is done simply by re-weighting: the formula used is

$$m^{-1}\sum_{j=1}^{m}\eta(\theta_j)w(\theta_j),$$

in which $w(\theta) = p(\theta \mid D)/q(\theta)$. The integral estimated by this sample average is

$$\int\eta(\theta)w(\theta)q(\theta)d\theta = \int\eta(\theta)p(\theta|D)d\theta = E\{\eta(\theta)|D\},$$

as required. The method works better when $q(.)$ is "close" to $p(. \mid D)$ and, in any case, $w(\theta)$ must be bounded as $|\theta| \to \infty$. In Rejection Sampling the values are discarded unless they satisfy a certain criterion designed to ensure that the retained values form a proper random sample from $p(\theta \mid D)$. For further details and other methods, see Morgan (1995) and Ripley (1987).

In recent years powerful general methods have been developed for hard cases where sampling directly from $p(\theta \mid D)$, or any suitable surrogate, $q(\theta)$, is unfeasible. These go under the general name of Markov chain Monte Carlo.

The idea is to hop around the parameter space in such a way that the points landed upon eventually resemble a random sample from $p(\theta \mid D)$. The hopping can be performed in various ways, either moving along individual coordinate directions one at a time, or in groups, or moving in the full dimensional space; the former method has come to be known, by a tortuous route, as Gibbs sampling. Each hop is made by generating a proposed move sampled from some chosen distribution; in some versions, the move can be rejected, in which case the chain remains at the current point which is then counted again in the sample. In "easier" cases, the chosen distribution can be the actual one, conditional on the non-moving components in the case of Gibbs sampling. In more realistic applications, such as those occurring in Competing Risks, the sampling distribution is chosen for convenience. In either case, the required convergence of the chain to the target distribution is achieved by suitable decisions on the moves. It is important to apply diagnostics to the output chain to check that it has covered the parameter space well.

For example, in a Metropolis random walk, a move is proposed from the current point and is then accepted or rejected, i.e., the chain either hops or stops. Acceptance of a proposed move is governed by a probability that favours moves to points of higher density, so such points figure larger in the resulting sample. In fact, this probability is cunningly contrived to produce just the right frequencies of sampled values in the long run. The proposed moves can be generated by sampling from any convenient distribution, the multivariate normal being a standard choice.

For fuller descriptions of all aspects of Markov chain Monte Carlo, including the generation of chains and convergence diagnostics, see the excellent publications listed in Section 2.3.3.

Bibliography

Aalen, O.O. (1975) Statistical Inference for a Family of Counting Processes. Ph.D. thesis, University of California, Berkeley.

Aalen, O.O. (1976) Nonparametric inference in connection with multiple decrement models. *Scand. J. Statist.* **3**, 15-27.

Aalen, O. (1995) Phase type distributions in survival analysis. *Scand. J. Statist.* **22**, 447-463.

Albert, J.R.G. and Baxter, L.A. (1995) Applications of the EM algorithm to the analysis of life length data. *Appl. Statist.* **44**, 323-348.

Ali, M.M., Mikhail, N.N., and Haq, M.S. (1978) A class of bivariate distributions including the bivariate logistic. *J. Multivar. Anal.* **8**, 405-412.

Allen, W.R. (1963) A note on the conditional probability of failure when hazards are proportional. *Oper. Res.* **11**, 658-659.

Altshuler, B. (1970) Theory for the measurement of competing risks in animal experiments. *Math. Biosci.* **6**, 1-11.

Andersen, P.K. and Gill, R.D. (1982) Cox's regression model for counting processes: a large sample study. *Ann. Statist.* **10**, 1100-1120.

Andersen, P.K. and Borgan, O. (1985) Counting process models for life history data: a review. *Scand. J. Statist.* **12**, 97-158.

Andersen, P.K., Borgan, O., Gill, R.D., and Keiding, N. (1993) *Statistical Models Based on Counting Processes*. Springer-Verlag, New York.

Anderson, T.W. and Ghurye, S.G. (1977) Identification of parameters by the distribution of a maximum random variable. *J. R. Statist. Soc.* B **39**, 337-342.

Arjas, E. and Greenwood, P. (1981) Competing risks and independent minima: a marked point process approach. *Adv. Appl. Probab.* **13**, 669-680.

Arjas, E. and Haara, P. (1984) A marked point process approach to censored failure data with complicated covariates. *Scand. J. Statist.* **11**, 193-209.

Arjas, E. (1985) Contribution to Discussion of Andersen and Borgan (1985).

Armitage, P. (1959) The comparison of survival curves. *J. R. Statist. Soc.* A **122**, 279-300.

Arnold, B. and Brockett, P. (1983) Identifiability for dependent multiple decrement/competing risk models. *Scand. Actuar. J.* **10**, 117-127.

Atkinson, A.C. (1985) *Plots, Transformations and Regression*. Clarendon Press, Oxford.

Bagai, I., Deshpande, J.V., and Kochar, S.C. (1989a) A distribution-free test for the equality of failure rates due to two competing risks. *Commun. Statist. Theory Methods* **18**, 107-120.

Bagai, I., Deshpande, J.V., and Kochar, S.C. (1989b) Distribution-free tests for stochastic ordering in the competing risks model. *Biometrika* **76**, 775-781.

Bancroft, G.A. and Dunsmore, I.R. (1976) Predictive distributions in life tests under competing causes of failure. *Biometrika* **63**, 195-217.

Barlow, W.E. and Prentice, R.L. (1988) Residuals for relative risk regression. *Biometrika* **75**, 65-74.

Barlow, R.E. and Proschan, F. (1975) Importance of system components and fault tree events. *Stoch. Proc. Appl.* **3**, 153-173.

Basu, A.P. (1981) Identifiability problems in the theory of competing and complementary risks — a survey. In *Statistical Distributions in Scientific Work*, Vol. 2, Taillie, Patil, and Baldesari, Eds., Reidel, Dortrecht, Holland, 335-348.

Basu, A.P. and Ghosh J.K. (1978) Identifiability of the multinormal and other distributions under competing risks model. *J. Multivar. Anal.* **8**, 413-429.

Basu, A.P. and Ghosh J.K. (1980) Identifiability of distributions under competing risks and complementary risks model. *Commun. Statist. Theory Methods* **A9**, 1515-1525.

Basu, A.P. and Klein, J.P. (1982) Some recent results in competing risks theory. In *Survival Analysis*, Vol. 2, J. Crowley and R.A. Johnson, Eds., IMS Monograph Series, Hayward, CA.

Beck, G. (1979) Stochastic survival models with competing risks and covariates. *Biometrics* **35**, 427-438.

Bedford, T. and Cooke, R.M. (2001) *Probabilistic Risk Analysis: Foundations and Methods.* Cambridge University Press, Cambridge.

Berkson, J. and Elveback, L. (1960) Competing exponential risks, with particular reference to the study of smoking and lung cancer. *J. Am. Statist. Assoc.* **55**, 415-428.

Berman, S.M. (1963) Note on extreme values, competing risks and semi-Markov processes. *Ann. Math. Stat.* **34**, 1104-1106.

Bernoulli, D. (1760,1765) Essai d'une nouvelle analyse de la mortalite causee par la petite Verole, et des avantages de l'inoculation pour la prevenir. *Mem. Acad. R. Sci.*, 1760, 1-45.

Boag, J.W. (1949) Maximum likelihood estimates of the proportion of patients cured by cancer therapy. *J. R. Statist. Soc.* B**11**, 15-44.

Boardman, T.J. and Kendell, P.J. (1970) Estimation in compound exponential failure models. *Technometrics* **12**, 891-900.

Borgan, O. (1984) Maximum likelihood estimation in parametric counting process models, with applications to censored survival data. *Scand. J. Statist.* **11**, 1-16 (Correction **11**, 275).

Breslow, N. and Crowley, J. (1974) A large sample study of the life table and product limit estimates under random censorship. *Ann. Statist.* **2**, 437-453.

Brooks, S.P. and Morgan, B.J.T. (1995) Optimisation using simulated annealing. *Statistician* **44**, 241-257.

Cheng, R.C.H. and Stevens, M.A. (1989) A goodness-of-fit test using Moran's statistic with estimated parameters. *Biometrika* **76**, 385-392.

Chiang, C.L. (1961a) A stochastic study of the life table and its applications: III. The follow-up study with the consideration of competing risks. *Biometrics* **17**, 57-78.

Chiang, C.L. (1961b) On the probability of death from specific causes in the presence of competing risks. *Proc. 4th Berkeley Symp.* **4**, 169-180.

Chiang, C.L. (1964) A stochastic model of competing risks of illness and competing risks of death. In *Stochastic Models in Medicine and Biology*, J. Gurland, Ed., University of Wisconsin Press, Madison, 323-354.

Chiang, C.L. (1968) *Introduction to Stochastic Processes in Biostatics.* John Wiley & Sons, New York.

Chiang, C.L. (1970) Competing risks and conditional probabilities. *Biometrics* **26**, 767-776.

Clayton, D. (1978) A model for association in bivariate life tables and its application in epidemiological studies of familial tendency in chronic disease incidence. *Biometrika* **65**, 141-151.

Clayton, D. and Cuzick, J. (1985) Multivariate generalisations of the proportional hazards model (with discussion). *J. R. Statist. Assoc.* A **148**, 82-117.

Collett, D. (1994) *Modelling Survival Data in Medical Research.* Chapman & Hall, London.

Cook, R.D. and Weisburg, S. (1982) *Residuals and Influence in Regression.* Chapman & Hall, London.

Copas, J.B. (1983) Plotting p against x. *Appl. Statist.* **32**, 25-31.

Cornfield, J. (1957) The estimation of the probability of developing a disease in the presence of competing risks. *Am. J. Public Health* **47**, 601-607.

Couper, D. and Pepe, M.S. (1997) Modelling prevalence of a condition: chronic graft-versus-host disease after bone marrow transplantation. *Statist. Med.* **16**, 1551-1571.

Cox, D.R. (1959) The analysis of exponentially distributed lifetimes with two types of failure. *J. R. Statist. Soc.* B **21**, 411-421.

Cox, D.R. (1962) *Renewal Theory.* Methuen, London.

Cox, D.R. (1972) Regression models and life tables (with discussion). *J. R. Statist. Soc.* B **34**, 187-220.

Cox, D.R. (1975) Partial likelihood. *Biometrika* **62**, 269-276.

Cox, D.R. and Hinkley, D.V. (1974) *Theoretical Statistics.* Chapman & Hall, London.

Cox, D.R. and Oakes, D. (1984) *Analysis of Survival Data.* Chapman & Hall, London.

Cox, D.R. and Reid, N. (1987) Parameter orthogonality and approximate conditional inference (with discussion). *J. R. Statist. Soc.* B **49**, 1-39.

Crowder, M.J. (1985) A distributional model for repeated failure time measurements. *J. R. Statist. Soc.* B **47**, 447-452.

Crowder, M.J. (1986) On consistency and inconsistency of estimating equations. *Econ. Theory* **2**, 305-330.

Crowder, M.J. (1989) A multivariate distribution with Weibull connections. *J. R. Statist. Soc.* B **51**, 93-107.

Crowder, M.J. (1990) On some nonregular tests for a modified Weibull distribution. *Biometrika* **77**, 499-506.

Crowder, M.J. (1991) On the identifiability crisis in competing risks analysis. *Scand. J. Statist.* **18**, 223-233.

Crowder, M.J. (1994) Identifiability crises in Competing Risks. *Int. Statist. Rev.* **62**, 379-391.

Crowder, M.J., Kimber, A.C., Smith, R.L., and Sweeting, T.J. (1991) *Statistical Analysis of Reliability Data.* Chapman & Hall, London.

Cutler, S. and Ederer, F. (1958) Maximum utilization of the life table method in analyzing survival. *J. Chron. Dis.* **8**, 699-712.

D'Alembert, J. (1761) Sur l'application du Calcul des Probabilites a l'inoculation de la petite Verole. *Opuscules* **II**, 26-95.

David, H.A. (1957) Estimation of means of normal populations from observed minima. *Biometrika* **44**, 282-286.

David, H.A. (1970) On Chiang's proportionality assumption in the theory of competing risks. *Biometrics* **26**, 336-339.

David, H.A. (1974) Parametric approaches to the theory of competing risks. In Reliability and Biometry: Statistical Analysis of Lifelength, F. Proschan and R.J. Serfling, Eds., SIAM, Philadelphia, 275-290.

David, H.A. and Moeschberger, M.L. (1978) *The Theory of Competing Risks*. Griffin, London.

Davis, T.P. and Lawrance, A.J. (1989) The likelihood for competing risk survival analysis. *Scand. J. Statist.* **16**, 23-28.

Desu, M.M. and Narula, S.C. (1977) Reliability estimation under competing causes of failure. In *The Theory and Applications of Reliability,* Vol. 2, C.P. Tsokos and I.N. Shimi, Eds., Academic Press, New York, 471-481.

Dinse, G.E. (1982) Nonparametric estimation for partially-complete time and type of failure. *Biometrics* **38**, 417-431.

Dinse, G.E. (1986) Nonparametric prevalence and mortality estimators for animal experiments with incomplete cause-of-death. *J. Am. Statist. Assoc.* **81**, 328-336.

Dinse, G.E. (1988) Estimating tumour incidence rates in animal carcinogeneity experiments. *Biometrics* **44**, 405-415.

Doganaksoy, N. (1991) Interval estimation from censored and masked failure data. *IEEE Trans. Reliab.* **40**, 280-285.

Doob, J.L. (1940) Regularity properties of certain families of chance variables. *Trans. Am. Math. Soc.* **47**, 455-486.

Doob, J.L. (1953) *Stochastic Processes*. John Wiley & Sons, New York.

Doyle, A.C. (1950) *Silver Blaze*. In *The Memoirs of Sherlock Holmes*. Penguin, London.

Ebrahimi, N. (1996) The effects of misclassification of the actual cause of death in competing risks analysis. *Statist. Med.* **15**, 1557-1566.

Efron, B. and Hinkley, D.V. (1978) Assessing the accuracy of the maximum likelihood estimator: observed versus expected Fisher information. *Biometrika* **65**, 457-487.

Elandt-Johnson, R. (1976) Conditional failure time distributions under competing risk theory with dependent failure times and proportional hazard rates. *Scand. Actuar. J.* **59**, 37-51.

Elandt-Johnson, R.C. and Johnson, N.L. (1980) *Survival Models and Data Analysis*. John Wiley & Sons, New York.

Elveback, L. (1958) Estimation of survivorship in chronic disease; the actuarial method. *J. Am. Statist. Assoc.* **53**, 420-440.

Farragi, D. and Korn, E.L. (1996) Competing risks with frailty models when treatment affects only one failure type. *Biometrika* **83**, 467-471.

Faulkner, J.E. and McHugh, R.B. (1972) Bias in observable cancer age and life-time of mice subject to spontaneous mammary carcinomas. *Biometrics* **28**, 489-498.

Finch, P.D. (1977) On the crude analysis of survivorship data. *Austr. J. Statist.* **19**, 1-21.

Fisher, L. and Kanarek, P. (1974) Presenting censored survival data when censoring and survival times may not be independent. In *Reliability and Biometry: Statistical Analysis of Lifelength*, F. Proschan and R.J. Serfling, Eds., SIAM, Philadelphia, 303-326.

Fix, E. and Neyman, J. (1951) A simple stochastic model of recovery, relapse, death and loss of patients. *Hum. Biol.* **23**, 205-241.

Fleming, T.R. and Harrington, D.P. (1991) *Counting Processes and Survival Models.* John Wiley & Sons, New York.

Flinn, C.J. and Heckman, J.J. (1982) New methods for analyzing structural models of labor force dynamics. *J. Econometrics* **18**, 115-168.

Flinn, C.J. and Heckman, J.J. (1983a) Are unemployment and out of the labor force behaviourally distinct labor force states?, *J. Labor Econ.* **1**, 28-42.

Flinn, C.J. and Heckman, J.J. (1983b) The likelihood function for the multistate-multi-episode model in 'models for the analysis of labor force dynamics.' In *Advances in Econometrics*, Vol. 3, R. Bassman and G. Rhodes, Eds., JAI Press, Greenwich, CT, 115-168.

Freund, J.E. (1961) A bivariate extension of the exponential distribution. *J. Am. Statist. Assoc.* **56**, 971-977.

Gail, M. (1975) A review and critique of some models used in competing risks analysis. *Biometrics* **31**, 209-222.

Gail, M., Lubin, J.H., and Rubenstein, L.V. (1981) Likelihood calculations for matched case-control studies and survival studies with tied death times. *Biometrika* **68**, 703-707.

Galambos, J. (1978) *The Asymptotic Theory of Extreme Order Statistics.* John Wiley & Sons, New York.

Gasemyr, J. and Natvig, B. (1994) Bayesian Estimation of Component Lifetimes Based on Autopsy Data. *Statistical Research Report No. 8*, Institute of Mathematics, University of Oslo.

Gelfand, A.E. and Smith, A.F.M. (1990) Sampling-based approaches to calculating marginal densities. *J. Am. Statist. Assoc.* **85**, 398-409.

Geyer, C.J. (1992) Practical Markov chain Monte Carlo. *Statist. Sci.* **7**, 473-511.

Gilks, W.R., Richardson, S., and Spiegelhalter, D.J. (1996) *Markov Chain Monte Carlo in Practice.* Chapman & Hall, London.

Gill, R.D. (1984) Understanding Cox's regression model: a martingale approach. *J. Am. Statist. Assoc.* **79**, 441-447.

Gill, R.D. and Johansen, S. (1990) A survey of product integration with a view towards application in survival analysis. *Ann. Statist.* **18**, 1501-1555.

Gill, P.E., Murray, W., and Wright, M.H. (1981) *Practical Optimization.* Academic Press, London.

Goetghebeur, E. and Ryan, L. (1990) A modified log rank test for competing risks with missing failure type. *Biometrika* **77**, 207-211.

Goetghebeur, E. and Ryan, L. (1995) Analysis of competing risks survival data when some failure types are missing. *Biometrika* **82**, 821-833.

Gompertz, B. (1825) On the nature of the function expressive of the law of human mortality. *Philos. Trans. R. Soc. London* **115**, 513-583.

Grambsch, P.M. and Therneau, T.M. (1994) Proportional hazards tests and diagnostics based on weighted residuals. *Biometrika* **81**, 515-526.

Greenwood, M. (1926) The errors of sampling of the survivorship tables. In Reports on Public Health and Statistical Subjects, No. 33, Appendix 1. Her Majesty's Stationary Office, London.

Greville, T.N.E. (1948) Mortality tables analyzed by cause of death. *Rec. Am. Inst. Actuaries* **37**, 283-294.

Greville, T.N.E. (1954) On the formula for the L-function in a special mortality table eliminating a given cause of death. *Trans. Soc. Actuar.* **6**, 1-5.

Gross, A.J. (1970) Minimization of misclassification of component failure in a two-component system. *IEEE Trans. Reliab.* **19**, 120-122.

Gross, A.J. (1973) A competing risk model: a one organ subsystem plus a two organ subsystem. *IEEE Trans. Reliab.* **22**, 24-27.

Gross, A.J., Clark, V.A., and Liu, V. (1971) Estimation of survival parameters when one of two organs must function for survival. *Biometrics* **27**, 369-377.

Guess, F.M., Usher, J.S., and Hodgson, T.J. (1991) Estimating system and component reliabilities under partial information on cause of failure. *J. Statist. Plann. Inf.* **29**, 75-85.

Gumbel, E.J. (1960) Bivariate exponential distributions. *J. Am. Statist. Assoc.* **55**, 698-707.

Guttman, I., Lin, D.K.J., Reiser, B., and Usher, J.S. (1995) Dependent masking and system life data: Bayesian inference for two-component systems. *Lifetime Data Anal.* **1**, 87-100.

Heckman, J.J. and Honore, B.E. (1989) The identifiability of the competing risks model. *Biometrika* **76**, 325-330.

Herman, R.J. and Patell, R.K.N. (1971) Maximum likelihood estimation for multi-risk model. *Technometrics* **13**, 385-396.

Hoel, D.G. (1972) A representation of mortality data by competing risks. *Biometrics* **28**, 475-488.

Holt, J.D. (1978) Competing risk analyses with special reference to matched pair experiments. *Biometrika* **65**, 159-165.

Hougaard, P. (1984) Life table methods for heterogeneous populations: distributions describing the heterogeneity. *Biometrika* **71**, 75-83.

Hougaard, P. (1986a) Survival models for heterogeneous populations derived from stable distributions. *Biometrika* **73**, 387-396.

Hougaard, P. (1986b) A class of multivariate failure time distributions. *Scand. J. Statist* **14**, 291-304.

Johnson, N.L. and Kotz, S. (1972) *Distributions in Statistics: Continuous Multivariate Distributions.* John Wiley & Sons, New York.

Kalbfleisch, J.D. and Sprott, D.A. (1970) Application of likelihood methods to models involving large numbers of parameters (with discussion). *J. R. Statist. Soc.* B **32**, 175-208.

Kalbfleisch, J.D. and Mackay, R.J. (1979) On constant-sum models for censored survival data. *Biometrika* **66**, 87-90.

Kalbfleisch, J.D. and Prentice, R.L. (1980) *The Statistical Analysis of Failure Time Data.* John Wiley & Sons, New York.

Kaplan, E.L. and Meier, P. (1958) Nonparametric estimation from incomplete observations. *J. Am. Statist. Assoc.* **53**, 457-481.

Karia, S.R. and Deshpande, J.V. (1997) Bounds for Hazard Gradients and Rates in the Competing Risks Set Up. *Technical Report 285*, Department of Statistics, University of Michigan, Ann Arbor.

Karn, M.N. (1931) An inquiry into various death-rates and the comparative influence of certain diseases on the duration of life. *Ann. Eugenics* **4**, 279-326.

Karn M.N. (1933) A further study of methods of constructing life tables when certain causes of death are eliminated. *Biometrika* **25**, 91.

Kay, R. (1986) Treatment effects in Competing Risks analysis of prostate cancer data. *Biometrics* **42**, 203-211.

Kimball, A.W. (1958) Disease incidence estimation in populations subject to multiple causes of death. *Bull. Int. Statist. Inst.* **36**, 193-204.

Kimball, A.W. (1969) Models for the estimation of competing risks from grouped data. *Biometrics* **25**, 329-337.

Kimball, A.W. (1971) Model I vs. Model II in competing risk theory. *Biometrics* **27**, 462-465.

King, J.R. (1971) *Probability Charts for Decision Making*. Industrial Press, New York.

Klein, J.P. and Basu, A.P. (1981) Weibull accelerated life tests when there are competing causes of failure. *Commun. Statist. Theory Methods* A **10**, 2073-2100.

Kochar, S.C. and Proschan, F. (1991) Independence of time and cause of failure in the multiple dependent competing risks model. *Statistica Sin.* **1**, 295-299.

Kodell, R.L. and Chen, J.J. (1987) Handling cause of death in equivocal cases using the EM algorithm. With rejoinder. *Commun. Statist.* A **16**, 2565-2585.

Krall, J.M. and Hickman, J.C. (1970) Adjusting multiple decrement tables. *Trans. Soc. Actuar.* **22**, 163-179.

Lagakos, S.W. (1978) A covariate model for partially censored data subject to competing causes of failure. *Appl. Statist.* **27**, 235-241.

Lagakos, S.W. and Louis, T.A. (1988) Use of tumour lethality to interpret tumorigenicity experiments lacking cause-of-death data. *Appl. Statist.* **37**, 169-179.

Lagakos, S.W. and Williams, J.S. (1978) Models for censored survival analysis: a cone class of variable-sum models. *Biometrika* **65**, 181-189.

Langberg, N.A. and Proschan, F. (1979) A reliability growth model involving dependent components. *Ann. Probab.* **7**, 1082-1087.

Langberg, N., Proschan, F., and Quinzi, A.J. (1978) Converting dependent models into independent ones, preserving essential features. *Ann. Probab.* **6**, 174-181.

Langberg, N., Proschan, F., and Quinzi, A.J. (1981) Estimating dependent life lengths, with applications to the theory of competing risks. *Ann. Statist.* **9**, 157-167.

Langberg, N. and Shaked, M. (1982) On the identifiability of multivariate life distribution functions. *Ann. Probab.* **10**, 773-779.

Larson, M.G. and Dinse, G.E. (1985) A mixture model for the regression analysis of competing risks data. *Appl. Statist.* **34**, 201-211.

Lawless, J.F. (1982) *Statistical Models and Methods for Lifetime Data*. John Wiley & Sons, New York.

Lawless, J.F. (1983) Statistical methods in reliability (with comments). *Technometrics* **25**, 305-335.

Lee, L. and Thompson, W.A., Jr. (1974) Results on failure time and pattern for the series system. In *Reliability and Biometry: Statistical Analysis of Lifelength*, F. Proschan and R.J. Serfling, Eds., SIAM, Philadelphia, 291-302

Lin, D.K.J. and Guess, F.M. (1994) System life data analysis with dependent partial knowledge on the exact cause of system failure. *Microelec. Reliab.* **34**, 535-544.

Lodge, D. (1990) *Famous Cricketers Series No. 15: W.G. Grace*. Association of Cricket Statisticians and Historians, West Bridgford, Nottingham, U.K.

Magaluri, G. and Zhang, C.-H. (1994) Estimation in the mean residual lifetime regression model. *J. R. Statist. Soc.* **56**, 477-489.

Makeham, W.M. (1860) On the law of mortality and the construction of annuity tables. *J. Inst. Actuaries* **8**, 301-310.

Makeham, W.M. (1874) On an application of the theory of the composition of decremental forces. *J. Inst. Actuaries* **18**, 317-322.

Marshall, A.W. and Olkin, I. (1967a) A generalized bivariate exponential distribution. *J. Appl. Probab.* **4**, 291-302.

Marshall, A.W. and Olkin, I. (1967b) A multivariate exponential distribution. *J. Am. Statist. Assoc.* **62**, 30-44.

Martz, H.F. and Waller, R.A. (1982) *Bayesian Reliability Analysis*. John Wiley & Sons, New York.

Matthews, D.E. (1984) Efficiency considerations in the analysis of a competing risk problem. *Can. J. Statist.* **12**, 207-210.

Meier, P. (1975) Estimation of a distribution function from incomplete observations. In *Perspectives in Probability and Statistics*, J. Gani, Ed., Applied Probability Trust, Sheffield, 67-87.

Mendenhall, W. and Hader, R.J. (1958) Estimation of parameters of mixed exponentially distributed failure time distributions from censored life test data. *Biometrika* **45**, 504-520.

Miller, D.R. (1977) A note on independence of multivariate lifetimes in competing risk models. *Ann. Statist.* **5**, 576-579.

Miyakawa, M. (1984) Analysis of incomplete data in a competing risks model. *IEEE Trans. Reliab.* **33**, 293-296.

Moeschberger, M.L. (1974) Life tests under dependent competing causes of failure. *Technometrics* **16**, 39-47.

Moeschberger, M.L. and David, H.A. (1971) Life tests under competitive causes of failure and the theory of competing risks. *Biometrics* **27**, 909-933.

Morgan, B.J.T. (1995) *Elements of Simulation*. Chapman & Hall, London.

Nadas, A. (1970a) On proportional hazard functions. *Technometrics* **12**, 413-416.

Nadas, A. (1970b) On estimating the distribution of a random vector when only the smallest co-ordinate is observable. *Technometrics* **12**, 923-924.

Nadas, A. (1971) The distribution of the identified minimum of a normal pair determines the distribution of the pair. *Technometrics* **13**, 201-202.

NAG (1988) *Library Manual Mark 13*. Numerical Algorithms Group, Oxford.

Nair, V.J. (1993) Bounds for reliability estimation under dependent censoring. *Int. Statist. Rev.* **61**, 169-182.

Nelson, W.B. (1970) Hazard plotting methods for analysis of life data with different failure modes. *J. Qual. Technol.* **2**, 126-149.

Nelson, W. (1982) *Applied Life Data Analysis*. John Wiley & Sons, New York.

Nelson, W. (1990) *Accelerated Testing: Statistical Models, Test Plans and Data Analysis*. John Wiley & Sons, New York.

Oakes, D. (1981) Survival times: aspects of partial liklihood. *Int. Statist. Rev.* **49**, 235-264.

Oakes, D. (1989) Bivariate survival models induced by frailties. *J. Am. Statist. Assoc.* **84**, 487-493.

Oakes, D. and Dasu, T. (1990) A note on residual life. *Biometrika* **77**, 409-410.

Peacock, S.T., Keller, A.Z., and Locke, N.J. (1985) Reliability predictions for diesel generators. *Reliability '85*, 4C/5/1–4C/5/12.

Pepe, M.S. (1991) Inference for events with dependent risks in multiple endpoint studies. *J. Am. Statist. Assoc.* **86**, 770-778.

Pepe, M.S. and Mori, M. (1993) Kaplan-Meier, marginal or conditional probability curves in summarizing competing risks failure time data? *Statist. Med.* **12**, 737-751.

Peterson, A.V. (1976) Bounds for a joint distribution function with fixed sub-distribution functions: application to competing risks. *Proc. Natl. Acad. Sci. U.S.A.* **73**, 11-13.

Peterson, A.V. (1977) Expressing the Kaplan-Meier estimator as a function of empirical sub-survival functions. *J. Am. Statist. Assoc.* **72**, 854-858.

Peto, R. (1972) Contribution to discussion of paper by D.R. Cox. *J. R. Statist. Soc.* B **34**, 205-207.

Peto, R. and Lee, P. (1973) Weibull distributions for continuous carcinogenesis experiments. *Biometrics* **29**, 457-470.

Peto, R. and Peto, J. (1972) Asymptotically efficient rank invariant test procedures (with discussion). *J. R. Statist. Soc.* **135**, 185-206.

Pike, M.C. (1966) A method of analysis of a certain class of experiments in carcinogenesis. *Biometrics* **22**, 142-161.

Pike, M.C. (1970) A note on Kimball's paper "Models for the estimation of competing risks from grouped data." *Biometrics* **26**, 579-581.

Pike, M.C. and Roe, F.J.C. (1963) An actuarial method of analysis in an experiment in two-stage carcinogenesis. *Br. J. Cancer* **17**, 605-610.

Prentice, R.L. and El Shaarawi, A. (1973) A model for mortality rates and a test of fit for the Gompertz force of mortality. *Appl. Statist.* **22**, 301-314.

Prentice, R.L., Kalbfleisch, J.D., Peterson, A.V., Flournoy, N., Farewell, V.T., and Breslow, N.E. (1978) The analysis of failure times in the presence of competing risks. *Biometrics* **34**, 541-554.

Proschan, F. and Sullo, P. (1974) Estimating parameters of a bivariate exponential distribution in several sampling situations. In *Reliability and Biometry: Statistical Analysis of Lifelength.* SIAM, Philadelphia, 423-440.

Rachev, S.T. and Yakovlev, A.Y. (1985) Some problems of the competing risks theory. In *5th Int. Summer School on Probability Theory and Math. Statist., Varna, 1985.* Bulgarian Academy of Science, Sofia.

Racine-Poon, A.H. and Hoel, D.G. (1984) Nonparametric estimation of the survival function when the cause of death is uncertain. *Biometrics* **40**, 1151-1158.

Reiser, B., Guttman, I., Lin, D.K.J., Guess, F.M., and Usher, J.S. (1995) Bayesian inference for masked system lifetime data. *Appl. Statist.* **44**, 79-90.

Ripley, B.D. (1987) *Stochastic Simulation*. John Wiley & Sons, New York.

Ross, G.J.S. (1970) The efficient use of function minimization in nonlinear maximum likelihood estimation. *Appl. Statist.* **19**, 205-221.

Sampford, M.R. (1952) The estimation of response-time distributions. II. Multi-stimulus distributions. *Biometrics* **8**, 307-369.

Schoenfeld, D. (1982) Partial residuals for the proportional hazards regression model. *Biometrika* **69**, 239-241.

Seal, H.L. (1954) The estimation of mortality and other decremental probabilities. *Skand. Aktuarietidskr.* **37**, 137-162.

Seal, H.L. (1977) Studies in the history of probability and statistics. XXXV. Multiple decrements or competing risks. *Biometrika* **64**, 429-439.

Self, S.G. and Liang, K. (1987) Asymptotic properties of maximum likelihood estimators and likelihood ratio tests under nonstandard conditions. *J. Am. Statist. Assoc.* **82**, 605-610.

Sen, P.K. (1974) Nonparametric Tests for Interchangeability Under Competing Risks. *Inst. Statist. Mimeo Ser. No. 905*, University of North Carolina, Chapel Hill.

Sethuraman, J. (1965) On a characterization of the three limiting types of the extreme. *Sankhya* A **27**, 357-364.

Shaked, M. (1982) A general theory of some positive dependence notions. *J. Multivar. Anal.* **12**, 199-218.

Slud, E.V. and Rubenstein, L.V. (1983). Dependent competing risks and summary survival curves. *Biometrika* **70**, 643-649.

Smith, A.F.M. (1991) Bayesian computational methods. *Philos. Trans. R. Soc. London Ser.* A **337**, 369-386.

Smith, A.F.M. and Roberts, G.O. (1993) Bayesian computation via the Gibbs sampler and related Markov chain Monte Carlo methods. *J. R. Statist. Soc.* B **55**, 3-23.

Takahasi, K. (1965) Note on the multivariate Burr's distribution. *Ann. Inst. Statist. Math.* **17**, 257-260.

Therneau, T.M., Grambsch, P.M., and Fleming, T.R. (1990) Martingale-based residuals for survival models. *Biometrika* **77**, 147-160.

Tsiatis, A. (1975) A nonidentifiability aspect of the problem of competing risks. *Proc. Natl. Acad. Sci. U.S.A.* **72**, 20-22.

Usher, J.S. and Hodgson, T.J. (1988) Maximum likelihood analysis of component reliability using masked system life-test data. *IEEE Trans. Reliab.* **37**, 550-555.

Usher, J.S. and Guess, F.M. (1989) An iterative approach for estimating component reliability from masked system life data. *Qual. Reliab. Eng. Inst.* **5**, 257-261.

Vaupel, J.W., Manton, K.G., and Stallard, E. (1979) The impact of heterogeneity in individual frailty on the dynamics of mortality. *Demography* **16**, 439-454.

Walker, S.G., Damien, P., Land, P.W., and Smith, A.F.M. (1999) Bayesian nonparametric inference for random distributions and related functions (with discussion). *J. R. Statist. Soc.* B **61**, 485-527.

Wang, O. (1977) A competing risk model based on the life table procedure in epidemiologic studies. *Int. J. Epidemiol.* **6**, 153-159.

Watson, A.S. and Smith, R.L. (1985) An examination of statistical theories for fibrous materials in the light of experimental data. *J. Mater. Sci.* **20**, 3260-3270.

Williams, D. (1991) *Probability with Martingales*. Cambridge University Press, Cambridge.

Williams, J.S. and Lagakos, S.W. (1976) Independent and dependent censoring mechanisms. *Proc. 9th Int. Biometric Conf.*, **1**, 408-427.

Williams, J.S. and Lagakos, S.W. (1977) Models for censored survival analysis: constant-sum and variable-sum models. *Biometrika* **64**, 215-224.

Wong, W.H. (1986) Theory of partial likelihood. *Ann. Statist.* **14**, 88-123.

Yashin, A.I., Manton, K.G., and Stallard, E. (1986) Dependent competing risks: a stochastic process model. *J. Math. Biol.* **24**, 119-164.

Zheng, M. and Klein, J.P. (1995) Estimates of marginal survival for dependent competing risks based on an assumed copula. *Biometrika* **82**, 127-138.

Index

A

accelerated life model, 16, 53, 139
actuarial approach, 111
adapted, 147
additive intensity model, 163
age-specific death rate, 2, 10
age-specific failure rate, 2, 10
at risk, 63, 70, 157

B

baseline hazard, 69, 70, 72, 77, 78
baseline survivor function, 79, 116, 117
bathtub hazard, 45, 102
Bayesian approach, 24, 33
Bayesian computation, 26, 171
Bernoulli trials, 59, 61, 89, 90, 91
bounds on distribution functions, 48, 86
breaking strengths, 6
Brownian motion, 45

C

cause-specific hazard function, 10
censoring, 19, 20, 62, 92, 109, 139, 158
censoring, informative, 140, 158
Chiang's proportionality assumption, 112
compensator, 150
complementary risks, 141
conditional expectation, 145
conditional odds, 86
confidence intervals, 24, 28, 71, 115
configuration, 85
conjugate directions, 169
conjugate prior, 25
copula, 97
counting process, 152
covariate, 13, 65
crude risks, 46, 111
crude survival functions, 38
cumulative hazard function, 59, 69, 107

D

demographic work, 111
density function, 1
dependence, 37
    measure, 50, 51, 88, 127
    negative, 39
    positive, 38
dependent risks, 119, 130
deviance residuals, 24
DFP algorithm, 169
diagnostics, 24, 73
difference operator, 85
discrete density function, 58
discrete failure times, 58, 76, 83, 117, 134
discrete hazard function, 58
distribution function, 1
distributions
    bivariate exponential, 39, 103, 121, 128
    bivariate Makeham, 103
    exponential, 4, 61
    exponential mixture, 5, 10, 14, 27, 103
    extreme value, 52
    gamma, 4, 25, 43, 61
    geometric, 59, 61, 84
    Gompertz, 103
    multinomial, 63
    multivariate Burr, 43
    multivariate Weibull, 44
    negative binomial, 61
    Pareto, 4, 26
    positive stable, 44
    waiting time, 59, 61

Weibull, 61, 132
Weibull mixture, 29, 102, 109
Doob decomposition, 150
Doob-Meyer decomposition, 150

E

empirical survivor function, 64, 68, 95
event history analysis, 143
explanatory variable, 13, 31, 32, 65, 137
external covariates, 78

F

failure configuration, 85, 123
failure pattern, 123
filtration, 147
Fisher's scoring method, 169
force of decrement, 2, 10
force of mortality, 2, 10
frailty, 42, 43
Freund bivariate exponential, 89, 137

G

Gibbs sampling, 172
goodness of fit, 22, 30, 73
Greenwood's formula, 65, 68
Gumbel bivariate exponential, 39, 103

H

hazard function, 2
hazard ratio, 33, 79
Hessian matrix, 24, 28
hypothesis tests, 24, 28, 71, 115

I

identifiability, 35, 101, 122, 130, 139–141
identified minimum, 141
immunity, 53
importance sampling, 171
independence, 37
independent risks, 48–50, 87, 88, 119, 123, 130, 135, 137
indicator function, 73
indicator process, 154
influence, 73
information, 24, 63, 64, 75, 94
instantaneous failure rate, 2, 10
integrated hazard function, 58, 68, 107, 138
integrated intensity function, 152
intensity function, 2, 10, 152
intensity process, 154
internal covariates, 78
interval-censored data, 66, 109
Ito process, 45

J

joint density function, 37
joint distribution function, 37
joint survivor function, 37

K

Kaplan-Meier estimator, 67, 68, 72, 106, 160
Kimball's assumption, 112

L

lack of memory property, 59
Laplace transform, 43
latent lifetimes, 37, 38, 52, 85, 119
Lehmann alternative, 23, 70
Levenburg-Marquardt compromise, 169
life expectancy, 16
life tables, 66
likelihood equations, 167
likelihood function, 19, 62, 65, 67, 72, 77, 92, 94, 155–157, 164, 167
likelihood ratio test, 24, 28, 29, 32, 33
logit model, 66, 77
log-linear model, 138
log-rank test, 163

M

Makeham assumption, 50, 51, 88, 103, 111, 112, 130, 140
marginal distribution function, 37
marginal hazard rate, 47
marginal intensity function, 47
marginal survivor function, 37
Markov chain Monte Carlo, 26, 33, 171
Marshall-Olkin bivariate exponential, 90, 93, 104, 128, 137
martingale, 148
counting process, 148
differences, 75, 76, 148
masked systems, 34
Match of the Day, 5

maximum likelihood, 21, 28, 30, 62, 63, 67, 94, 106, 167
maximum partial likelihood estimator, 73, 74, 114
mean residual life, 16, 23, 73
measurable, 145
Metropolis random walk, 33, 172
mixed survival distributions, 59
mixture models, 91
moment generating function, 43
Moran's test, 24, 30
multiple causes, 85
multivariate survival distribution, 37

N

Nelson-Aalen estimator, 69, 159, 161
nervous nineties, 104
net risks, 46, 111
net survival functions, 38
Newton-Raphson, 168
non-parametric estimation, 63, 68, 94, 106, 159, 161
non-repairable series systems, 4
normal deviate, 28, 32, 33

O

orthogonal, 64, 73, 78, 117, 140, 151, 169
outlier detection, 73

P

parallel systems, 141
parametric estimation, 62, 93
partial likelihood, 70, 74, 78, 114, 117, 164
partial score function, 74, 164
posterior density, 26
posterior distribution, 25, 171
posterior mean, 26
predictable, 149
predictable covariation, 151
predictable variation, 150, 160
prediction, 25, 26
probability integral transform, 23, 73
probability measure, 145
probability space, 145
product integral, 153
product limit estimator, 68, 72
proportional hazards, 12, 15, 32, 50, 69, 76, 79, 84, 88, 113, 121, 138, 139, 163, 164
proportional odds, 16, 77, 117
proxy model, 123, 124, 130, 135, 137
p-value, 28

Q

quasi-Newton methods, 169

R

regression model, 13, 31, 32, 66, 69, 137, 163
relative risk, 33
reliability, 4, 48, 102, 123
residuals, 23, 24, 30, 73
  deviance, 164
  martingale, 73, 164
  martingale transform, 165
  Schoenfeld, 165
  score, 165
risk-removal model, 53
risk set, 66, 70, 76, 94, 114

S

sample path, 147, 163
sample space, 144
score function, 24, 167
semi-parametric estimator, 70, 164
series systems, 141
sigma field, 144
steepest descents, 168
stochastic differential equation, 45
stochastic integral, 151, 160, 162, 164
sub-conditional-odds ratios, 86
sub-density function, 5, 84
sub-distribution function, 4, 83
sub-hazard function, 10, 20, 84
sub-mrl function, 17
sub-survivor function, 4, 83
survivor function, 1

T

ties, 72, 78, 123, 129, 132
time-dependent covariates, 78, 163
total time on test, 22

U

unbiased estimating equation, 75
unidentified minimum, 141
usual regularity conditions, 75

V

variable metric, 169

W

Wald test, 28, 99, 116
Wiener process, 45